VALDEZ GOLD RUSH TRAILS OF 1898-99

Jim and Nancy Lethcoe

Prince William Sound Books
Valdez, Alaska
♻ Printed on recycled paper

Cover credits: Front cover: From the Geological Reconaissance map of a part of The Copper River and Adjacent Territory, Alaskan Military Expedition, 1898, in Glenn and Abercrombie, Explorations in Alaska, 1899. Courtesy of the Valdez Museum and Historical Archives.

Back cover: Top Photograph, Neal Benedict from Messer Collection, courtesy of the Cook Inlet Historical Society. Bottom photograph: Panorama of Valdez, 1903 by Miles Brothers. From the Crary Collection, courtesy of The Anchorage Museum of History and Art. B62.1.811A.

ISBN-1-877900-05-2

Acknowledgments

As writers on local history, we would not have been able to research this book without the support of many people in our community. For many years, the Valdez City Council has financially supported The Valdez Museum and Historical Archives. We thank them for their remembrance of Valdez's past and appreciation of what that heritage means to both old-timers and new-comers in the fostering of a sense of community identity and the feeling of belonging.

We wish to express our appreciation to Joe Leahy, the director of the Valdez Museum, for helping us to locate many of the sources cited in this work.

We extend special thanks to Dorothy Clifton who has dedicated many years of her life to preserving the history of our town. She has spent many hours searching Valdez Historical Society materials for us.

A number of gold rush participants left diaries, journals, memoirs, letters and scrapbooks. Many of these remain unpublished, but the heirs generously granted us permission to use them. We especially wish to thank Klondike Press for permission to quote from *The Alaska Gold Rush Letters of Leroy Stewart Townsend: 1898-1899*, Copyright © 1995 by Peggy Jean Townsend and Klondike Press, publication in press, due in 1997. The Townsend letters add great depth to our understanding of life along the Valdez Gold Rush Trail. We encourage our readers to consult this fascinating work on publication.

Mr. Lestor Bourke kindly granted us permission to quote from Joseph Bourke's extensive *Journal*, notes and *Scrapbook*. Joe Bourke ultimately made Valdez his home and contributed greatly to the town's development.

Mrs. William E. Martin, the widow of Treloar's grandson, most graciously granted us permission to quote from William E. Treloar's wonderfully descriptive memoirs.

For the use of George C. Hazelet's extensive diaries, we extend our thanks to Calvin Hazelet and Elizabeth Towers. Elizabeth Towers' new book, *Icebound Empire*, portrays the saga of the development of the copper claims mentioned here.

The Alaska State Library granted us permission to quote from Neal D. Benedict's *The Valdes and Copper River Trail, Alaska*, while The Anchorage Museum of History and Art gave us permission to use documents and photographs from the Crary Scrapbooks. We thank them both.

Most importantly, the many relatives of participants in the Valdez Gold Rush Trails of 1898-99 deserve thanks for placing their family documents in public collections. Without them, the stories of the Valdez gold rush trails, the triumphs and hardships of those who participated, would have been lost.

Table of Contents

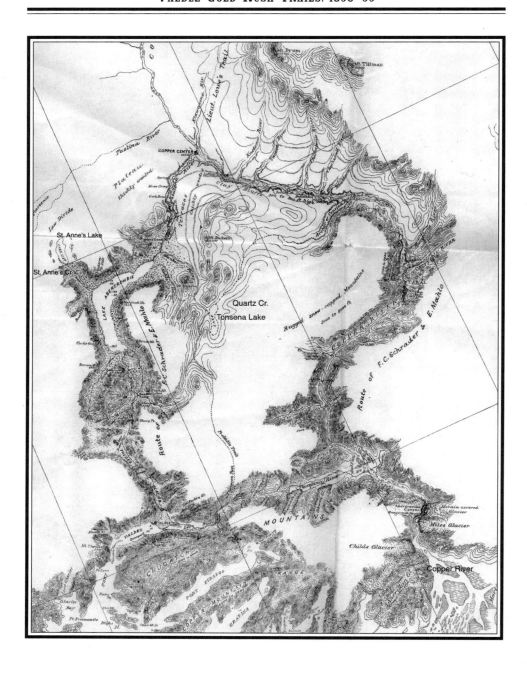

General map of the area showing what was known of the routes to the Copper River basin in 1898. From the Geological Reconaissance map of a part of The Copper River and Adjacent Territory, Alaskan Military Expedition, 1898, in Glenn and Abercrombie, Explorations in Alaska, 1899. Courtesy of the Valdez Museum and Historical Archives.

Chapter 1
The Gold Fever of '98

The Steamer *Valencia* arrives in Port Valdes

Despite the howling wind and the sub-zero chill factor, hundreds of anxious prospectors crowded the decks of the *Valencia* as she steamed through the Narrows and into Port Valdez on March 18, 1898. Bracing themselves against the icy blasts, they strained for a first glimpse of the glacier that was extolled as a highway to the fabled gold fields of the Copper River region. As they rounded Entrance Point, the scene that greeted them was one of awesome beauty yet harsh portent. Bathed in the dazzling light of the late winter sun, jagged snow-clad peaks glistened a full 5,000 feet above the blue and white, storm-whipped waters of Prince William Sound. Snow reached from these lofty heights down to the very water's edge. Snow plumes trailed in the lee of the towering peaks while furious gusts issuing from the glacier's terminus swept sheets of spray, sheathing the iron-hulled steamer in ice. The wind howled in the ship's rigging like a thousand screaming demons from hell. In Seattle they had been promised there would be a dock at Port Valdez. However, passenger Joseph Bourke surveying the scene observes: "Arriving at Valdez, we found no wharf nor storehouse or any other convenience, nothing in fact but a snow bank (Bourke, p. 4)." Four of the ship's passengers were so intimidated by the scene that they immediately booked passage back to Seattle on the return voyage.

Indeed, the outbound voyage of the *Valencia* had been a hard one on the passengers, but lucrative for the ship's new owners. The Pacific Steam and Whaling Company had purchased the steel-hulled, sailing vessel in New York and hastily converted it into a passenger steamship. The company had only recently entered the lucrative, gold rush passenger trade. As operators of the Prince William Sound salmon cannery at Orca (near present day Cordova), it began promoting the Valdez Glacier route as an easy, All-American trail to the interior gold fields. This made good economic sense — for the shippers at least. Rather than run empty vessels to Alaska to fetch the salmon pack, the company could now carry paying passengers.

Bound round Cape Horn from the east coast to Seattle, the *Valencia* called at various ports along its route boarding eager stampeders for the Klondike and Alaska gold fields. By March 6, 1898, when the *Valencia* departed Seattle for Port Valdez, she was grossly overloaded carrying over 640 passengers, plus a large number of horses and cattle, several oxen and a number of dogs. The crowded, second-class accommodations between decks consisted of straw mattresses laid side by side over hastily hewn, rough planks. The food was predictably awful. Joe Bourke remarks

1

The Pacific Steam Whaling Co. purchased the old sailing vessel Valencia *and converted her to steam. The* Valencia *picked up her first passengers on the east coast, then after rounding Cape Horn, boarded more passengers in San Francisco and Seattle before heading to Valdez. Gold rushers making this voyage were extremely critical of the company.*

Photograph by Neal Benedict from the Messer Collection, courtesy of the Cook Inlet Historical Society.

that the *Valencia* was "better adapted for carrying cattle than human beings." Bourke further complains: "The Pacific Steam Whaling Co of Seattle should be shown up in their proper light in the East with people cautioned to keep clear of them. It is a complete swindle. People pay seventy dollars for what they term second class passage, and when they are on the boat are treated like hogs (Bourke, *Journal,* p. 1)."

Crossing the Gulf of Alaska, the heavily laden steamer encountered an awful three-day storm. The frightened passengers rolling in their stuffy, dark, dank and crowded quarters suffered acutely from seasickness. The livestock on deck were tossed about so badly that many were crippled and had to be thrown overboard. The advertised five day trip occupied a full eleven excruciating days. Thus, it is little wonder that the captain on arriving in Port Valdes had a major revolt on his hands. Because of the severe winds, the captain anchored the steamer on the south side of the bay at the protected anchorage of Swanport. The prospectors, who had no means of conveying themselves or their supplies the four miles across the open water to the head of the glacier trail, forced the captain at gun point to order tenders from the company's Orca cannery. The cannery tenders then transported the prospectors and their year's supply of goods to the tent town on the opposite shore known as "Copper City." Some of the new arrivals perceived immediately that maintenance of law and order on this remote frontier was bound to create problems; even before stepping ashore, they duly elected a constable and deputies.

The Historical Setting

From our remote position in time, it requires an extreme exercise of historical imagination to understand what possessed between three and four thousand men and women to suddenly abandon their families, farms, jobs, professions; to endure the hardships and privations of a long sea voyage; land in a remote, often harsh, environment devoid of all civilized conveniences; cross a forbidding glacier; shoot the rapids of wild rivers and trek hundreds of miles through an unexplored wilderness — all in quest of the elusive promise of gold. Whatever their motivations — be it an all too human hodge-podge of folly and courage, gullibility and optimism, greed and self-sacrifice — their quest made a difference. Their efforts led not only to the exploration and opening of much of interior Alaska, but also to the establishment of the city of Valdez which would become the early metropolitan, legal and communications center for all of south-central Alaska; the major supply and transportation center for interior Alaska; and finally the terminus of the trans-Alaska pipeline.

To understand what brought the gold seekers of the *Valencia* to Port Valdez on that windy March day, we must look to the tenor of the times and to the factors that precipitated the great Klondike gold rush of '98. The so-called "gay '90s" were really not all that gay for most Americans. The gaiety of the period was not so much a celebration of life as a frantic escapism from the bleak sense of hopelessness that gripped an entire nation at the end of the nineteenth century.

By the late 1800s, the population of the United States was largely urban while its traditions and values remained decidedly rural and even frontier. The era was nurtured on the exploits of living ancestors who in a series of successive gold rushes to California, Montana, and Colorado had pushed the American frontier to the very brink of the Pacific. The preceding era had been one of unlimited opportunity based on apparently inexhaustible resources and limitless westward expansion. It was a time when the hard working, self-sufficient and resourceful individual willing to take the necessary risks could rise above the most humble of origins and end his life a wealthy man. This, after all, is still the stuff of the American Dream.

The final railroad was pushed through to the west coast in 1869 signaling the end of an era. Here, on the shores of the Pacific Ocean, the track ended abruptly. There would be no more westward expansion, no more unlimited opportunity. The image of two lonely tracks poised on the brink of a seemingly empty and endless ocean fittingly sums up the hopelessness with which many men viewed their lives at the end of the nineteenth century.

The mighty railroads themselves at the end of this period of rapid expansion suddenly began to experience financial difficulties. The Philadelphia and Reading Railroad failed in January of 1893. It was soon followed by the Erie, Union Pacific, and Northern Pacific. That such enterprises, which had so recently tamed half a continent, should fail seemed unthinkable. The financial markets panicked and the American economy rapidly lost its sinew. Money and credit became so tight that by the end of 1895 many of the nation's most successful businesses were forced into bankruptcy. Hundreds of thousands of workers were turned out onto the streets. Wealthy Ohio reformer, Jacob Coxey, and "Coxey's Army" of 400 marched on Washington to seek relief while clashes between big business and labor further immobilized the economy.

The panic of 1893 and the ensuing depression had a devastating effect on the American psyche. People began to hoard gold. To make matters worse, the U.S. Government clung stubbornly to the gold standard insisting on paying its debts in that currency. U.S. Gold reserves were so drained by 1896, that there was insufficient gold for adequate coinage thus further reducing the money supply. Gold was on everybody's mind, and the stage was set for the events of the summer of 1897.

GOLD! GOLD! GOLD! GOLD!

Sixty-eight Rich Men on the Steamer *Portland*

Stacks of Yellow Metal !

The Klondike and Copper River Gold Rush

These headlines of the *Seattle Post-Intelligencer* of July 17, 1897 suddenly broke the dams of despair. For the next nine months, scores of thousands of people from all walks of life and from all parts of the nation boarded trains for the west coast ports of Seattle and San Francisco. Here hastily salvaged hulks waited to transport them to the gold fields of the Klondike, the Copper River area and later Nome. Most were urban dwellers who knew absolutely nothing about conditions in Alaska or about wilderness survival, or about geology and gold mining; but news of the Klondike strike had touched a national nerve. Over fifty thousand people were inspired to leave businesses, jobs, homes, family and friends risking everything in hopes of striking it rich in the Klondike gold fields. The popular perception was that all one had to do was make the arduous journey to the Klondike and once there scoop up the gold which lay waiting on practically every gravel bar. Ignorance of Pacific Northwest geography was so great that most did not even know that the Klondike was in Canada and, in fact, used the term to vaguely refer to interior Alaska and Canada.

Certainly, most were victims of their own gullibility. Undoubtedly, the desire to better their economic circumstances during difficult times led many to dream unrealistic dreams of instant wealth and success. However, the American press was largely responsible for successive waves of misinformation that swept across the country. The telegraph, teletype and transoceanic cables had all recently revolution-ized communications so that rumors published in the S*an Francisco Chronicle* would appear almost simultaneously in the *New York Times* and the *Times of London*.

No matter how far-fetched or unreliable the rumors, the nation and world were hungry for news of the Klondike. The very atmosphere was alive with this magic word. If a merchant wanted to sell his wares, he merely need attach the word "Klondike" to his product and it would sell-out immediately. One could buy Klondike glasses, Klondike boots, Klondike sleeping bags — even Klondike soup.

Outfitting stores across the nation soon depleted their stocks of outdoor gear while woolen mills could not keep pace with the demand for woolen blankets. It seemed every able-bodied person was heading for the Klondike to gather up the gold.

Reliable information about Alaska and the Klondike was very much in demand and short in supply. However, this did not deter newspapers of the day which were fiercely competitive. Soon most realized that sensationalism increased circulation. Unnamed sources were often quoted, and editors were not very critical in chasing down the sources of rumors that might have great reader appeal. Returning prospectors, a promoter from the Seattle Chamber of Commerce and an employee of a shipping company were all accepted uncritically as authorities and reliable sources. For example, the reporter for the *Seattle Post-Intelligencer* article headlined above enthuses: "Conservative men who have just been in the country (the Klondike) . . . admit that all the fields in the vicinity of the Clondyke have been taken, but every river in Alaska is, in their judgment filled with gold which can be secured if the men are willing to risk the hardships (7/17/97)." By the winter of 1897, maps (many of them bogus) and Klondike guidebooks, by supposedly reputable authorities, suddenly appeared on the newsstands.

Picking the Best Route and Destination: The Newspapers

One of the first pieces of information of interest to the would-be gold seeker was his choice of a route. Experts agreed that there were really only two major routes to the gold fields. The first, the so-called "rich man's route," was by steamer or sailing vessel across the Gulf of Alaska and up the Bering Sea to the mouth of the Yukon River at St. Michaels. From here a steamer would carry prospectors 1700 miles up the Yukon River to Dawson. The disadvantage of this route was that it necessitated a long sea voyage (3000 miles) across notoriously stormy seas; moreover, the Bering Sea and the Yukon River were frozen the greater part of the year. The second, shorter, "poor man's route," led up the inside passage of British Columbia and Southeast Alaska to either Dyea or Skagway. From one of these ports, the prospector had to haul his entire outfit (often weighing between 1500 and 2000 lbs) up and over either the Chilkoot or White Pass. Once over these formidable barriers, he had to lug and float his gear 600 miles down the Yukon River to the Klondike. On entering Canadian territory, the American prospector was required to pay duty on his entire outfit. The Canadian Government further insisted that anyone entering the country by this route must bring with him a year's worth of food and supplies. Prospectors reasoned that surely there must be an easier, All-American route. At least one guidebook, the San Francisco newspapers, and a shipping company employee claimed there was — the Copper River Route.

Noted Klondike authority, A. C. Harris, in his hastily published 1897 guidebook, *The Klondike Gold Fields,* describes a new All-American, "back door route" to the gold fields:

> J. M. C. Lewis, a civil engineer, has proposed to the Interior Department, at Washington, a route from the mouth of the Copper River, by which he says the Klondike may he reached by a journey of a little over 300 miles from the coast, a great saving in distance over the other mountain routes. He says the trail could be opened at small expense.
>
> The route which he proposes will start inland from the mouth of the Copper River, near the Miles Glacier, twenty-five miles east of the entrance to Prince William Sound. He says the Copper River is navigable for small steamers for many miles beyond the mouth of its principal eastern tributary. . . .(Harris, p. 159).

Newspapers quickly picked up this advice and spread it across the nation. However, anyone familiar with the government reports of Lieut. W. R. Abercrombie's abortive attempt to ascend the Copper River in 1884 would have known of the shallow delta with its frustrating sloughs and braided streams, the dangerous icebergs calving from Childs and Miles glaciers, and the impassable Abercrombie Rapids. And, although Lieut. Allen's handpicked team of three successful ascended the Copper River in 1885, the journey was extremely arduous with the men nearly starving to death. The newspaper recommendations were pure nonsense.

Most people, however, read newspapers and not government reports thus leading to the following imaginative but absurdly unrealistic early *Prospectus* of the Connecticut and Alaska Mining and Trading Association.

> We propose to start for the mouth of the Copper River leaving Seattle about Feb. 1st. We will sail up the river as far as the rapids which are about 100 miles [sic 35 miles] up from the mouth, we shall then launch the steam launch, load it with provisions for a two weeks trip, rope, blocks, arms, ammunition, the boreing machine and ten men and start up through the rapids, if the boat will not make any headway against the rapids we will send some of the men up the bank till they get above the rapids, then they can make fast one end of the rope which they will take with them to a tree or spur of rock and tie the other end to a log throw it in the river where the swift current will carry it down to the waiting men in the launch, who will speedily haul themselves up through the rapids. Once over them we have 200 miles of good navigable river, we will steam right ahead till we find the miners who have gone in there ahead of us and if they have found gold in good paying quantities we will locate near them and stake out claims . . . (Crary Scrapbook, Vol. 1. p. 112).

Government Reports

Reading government reports, however, could be equally misleading. In 1884, the military had dispatched young Lt. W. R. Abercrombie to the Copper River with instructions to ascend the river to investigate rumors of hostile natives in the upper Copper River region who had reportedly exterminated an earlier Russian expedition. The army desired to avoid any further conflicts between Copper River Indians and the white Americans whom they knew would inevitably settle the interior. Abercrombie did not arrive at the mouth of the Copper until June 21 when the river was flooding and choked with glacier ice. After spending the rest of the summer in a futile attempt to force his way up the raging river, Abercrombie in early September made his way back to the Alaska Commercial Company trading post at Nuchek near the entrance to Prince William Sound. Here, he learned of another route to the upper Copper River region from the head of the northeastern arm of the Sound — Port Valdes. His official 1885 report reads:

> I had been informed by the upper river Indians and those on the coast that a shorter route via Port Valdez led over the mountains to a lake the outlet of which ran into the Copper River below Chettyna. Its history related by the old Russian, is as follows: Some years ago, probably twenty or thirty, the portage was used entirely by the Upper River Indians, who came down the Copper River to the stream heading in the lake, which not being previously named or visited by white men, is designated Lake Margaret. Up this they traveled to the lake, hence to the foot of the passage. Here they left their bidarra and packed their furs over to salt water, where bidarras were furnished by the Chugachimutes for the voyage to Port Etches (Abercrombie, 1885, p. 389).

Abercrombie started from Nuchek at Port Etches for Port Valdes on September 10th with Lieut. Brumbach and a Russian Creole in a large canoe manned by three natives. At Port Fidalgo they pick up the Creole son of Plutinoff, a Russian experimental rancher, to guide them across the glacier. Arriving two days later, Abercrombie and his guides explored the glacier route.

> On the morning of Sept. 12 the expedition started up the trail at daylight, two half-breeds in the lead. Each had a blanket and a dried fish. Myself and Brumbach carried a blanket and overcoat, rolled up in which was some frying-pan bread for supper and for breakfast to partake of on the lake.

> After an arduous climb for six hours we reached the glacier in the following order, viz. half-breed Russian, son of Plutinoff, myself, Nicholi Nicolsky, half-breed from Nuchek, and Lieutenant Brumbach. After traveling an hour or more the fog settled down and we could not see more than 50 yards. About this time a call from the rear was heard from Lt. Brumbach. . . . it was found that he had taken with violent cramps in the muscles of his legs. On returning to him, he found him lying in the snow. I agreed to go far enough to see the lake and locate its outlet and return to him. The snow was very soft and

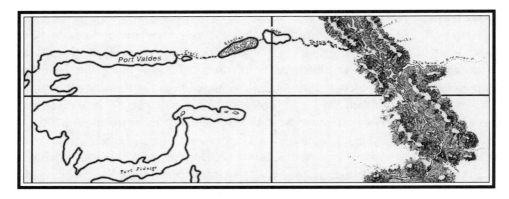

Allen's 1885 map included information from Abercrombie's 1884 alledged ascent of the Valdez Glacier Trail. Note that the glacier is shown running east/west when, in fact, it flows north/south. The map correctly indicates a river flowing from the glacier's inland side, but it is the Klutina River not the Tasnuna. When this and similar erroneous maps were published in newspapers during the winter of 1897-98, many prospectors decided to pursue the Valdez Glacier Trail to the reputed Copper River Gold Fields. From Abercrombie, Compilations.

rotten and the guide fell through. Had I not been close at hand to drag him out, the guide would have lost his life. No one could see the bottom of the fissure, but water could be heard rushing past in the darkness below. After locating the lake and its outlet, I returned and found Lieutenant Brumbach still on the snow. The guides informed me that the snow was too thin for packing, but that in two months the passage could be made safely. On returning to camp, which was reached after dark, I found my feet partly frozen. . . (Abercrombie, 1885, pp. 391-392).

Abercrombie sums up the result of his reconnaissance of the glacier portage thus:

Rounding a sharp turn in the inlet, the portage lies between two mountains, the valley being filled with a large glacier. The estimated altitude of the highest point of this portage is about 2500 ft. And from the base of the mountains on west to the lake on the east is about 15 miles (p. 391).

Much to the consternation of later prospectors, who knew of the above report or had maps based on it, much of the information was hopelessly and dangerously false and incorrect. First, when they traversed the 4-1/2 miles to the glacier's terminus on a practically level outwash plain, it was difficult to understand the lightly loaded Abercrombie's "arduous six hour climb." Secondly, whereas Abercrombie describes the glacier as trending in an east-west direction, their compasses definitely indicated a direction of south to north. In addition, Schrader (1898) found the altitude of the summit to be 4800 feet rather than only 2500, and the distance to the lake is 46 miles not a mere 15. Furthermore, if Abercrombie took six hours merely to reach the glacier terminus, one cannot believe he covered the remaining 40 miles round trip to the summit and back to tidewater in the time allowed. Finally, Klutina Lake and its outlet are obscured from view from the summit of Valdez Glacier by an intervening mountain ridge; so there is no way Abercrombie could have located the lake from the summit. It is interesting to note that when Abercrombie makes his alleged final ascent to the summit, the only witness capable of questioning his report, Lt. Brumbach, was left behind with cramps. It would be fairly safe to assume that his Creole Russian guides would never read his report. Although Lt. Abercrombie was later to play a significant role in the development of trails to the interior through Valdez, we must concur with Louis Guiteau's evaluation:

> . . . this little known route to the interior of Alaska [the Valdez Glacier trail] came into being as a result of a series of tragic errors, of government misinformation and a map made, years before, by an army officer from his imagination, made to forestall any criticism by his superiors, of his failure to accomplish his mission which had been to locate such a route (Louis Guiteau, *Death Trail*, p. 3).

Because most of the Klondike region had already been staked by 1897, newspapers feeding the gold fever of the period were anxious to ferret out any rumors of rich finds elsewhere in the north — especially in Alaska. Although Lt. Allen mentioned that he had found little indication of gold on his expedition up the Copper River, rumors circulated that the upper reaches of the Copper River were rich in gold. In February of 1898, engineer, Alexander MacDonald, in charge of the government's geological survey in Alaska for six years is reported in the Sitka newspaper to have proclaimed: "the Copper river was the easiest route to an area of 200 square miles, which he considered far richer than the Klondike." He is quoted as claiming "the gold fields on the American side are probably richer and far more extensive than the Klondike fields, which are on the British side. . . . There is gold in Alaska all the way to the coast (*The Alaskan*, 2/28/1898)." Ella Lang Martinsen in her book *Black Sand and Gold* reports a rumor that "an old trader named [Tom] Olson from Valdez on Cook Inlet [*sic*] reported Tanana Indians arriving with large gold nuggets that they brought from across Mentasta Pass (Martinsen, p. 307)."

Probably, the most intriguing rumor of Copper River gold is one reported by surveyor, Addison Powell, in his wonderful account of the Valdez gold rush, *Trailing and Camping in Alaska.* Powell recalls a February 1898 meeting with old-timer, Captain I. N. West in a San Francisco hotel room. Desiring to engage Powell to survey his gold discovery in Alaska, West confides in him a rich strike he made in the 1880s on the upper reaches of the Copper River. He claims to have been guided there by nine Yakutat Indians across the St. Elias Range by way of a glacier located half way between the Malaspina and Bering Glaciers. After an extensive survey of the upper Copper River region, he claims to have discovered gold on the Chistochina

> Well, we descended [sic] the Copper as far as the Chistochina River. There I discovered fine gold coming down that river, so we hid our raft in the

brush, and spent two weeks up at the source . . . . Now, let me tell you something about the discovery. The Indian found the first nugget, which he picked up with his hand. I then washed out considerable gold with my pan. I had to take it down nearly a quarter of a mile to where there was a little water at the junction of another creek. We found, on the mountain-side, a very rich pocket, and the Indian carried the gravel down in a sack. I continued to wash until I had panned out about six hundred dollars. The only thing that bothers me is the scarcity of water, but of course that is more plentiful in summer-time, as it was very late in the fall when we were there. Gold! (Powell, *Trailing and Camping in Alaska,* pp. 6-7).

West continues:

We returned as far as the Klutina River. From there we ascended to a lake which is about twenty miles long. Of course, there is no lake marked on the maps, but it is there, all the same. On the west side of this lake I am going to attempt to ascend a creek that leads over towards the Chistochina. I believe I can get through that way. From that lake we crossed the glacier over to Valdez Bay. We hoped to find a trading post there, but there was none, and we built a raft and floated with the tide, about twenty-five miles, to where we arrived at an Indian town. We were taken from here in bidarkies to Nutchek. The *Jeanie* soon arrived there on her return from a whaling cruise, and on it I returned to San Francisco (Powell, *Trailing and Camping in Alaska,* p. 7).

Powell was convinced enough by this story that he traveled to Valdez in June to survey West's claims. West arriving earlier had already traversed the glacier into the interior where he was taken ill. He was brought back out to Valdez and, unfortunately, departed on a steamer three days before Powell's arrival.

If West's account is true, he was among the first white men to cross Valdez Glacier. Men, Powell met later on the Valdez trail, relegated West's apparently oft-repeated account to the trash heap of rumor: "Know that old humbug! Well, I reckon I do! If it had not been for that old scoundrel we should not be camped here." Powell continues, "I found numbers of them who could not say enough against West. They pronounced him a humbug and a fraud who was working for the transportation companies." (Powell, *Trailing and Camping in Alaska,* p. 45).

Powell, however, becomes less skeptical. On his travels through the interior, he finds West's geography, at least in one instance, to be more accurate than Lt. Allen's and believes that West's discovery was the strike subsequently made by Hazelet and Meals and others on the Chistochina. Later, he checks with an officer of the whaler, *Jeanie*, who confirms West's story that he was picked up at Nuchek with gold from the Copper River.

About the same time as Powell's meeting with West, an interview with an old Alaskan trapper by the name of Bayles appeared in the *Seattle Times.* Bayles is quoted as saying, "Today the Klondike country is claiming all the attention but the near future will demonstrate that the rich gold region comprises a territory fifty times as large as what is now known as the Klondike country. The Copper River country alone, with its tributaries, is over 250 miles long, by 150 miles broad, which means an area of 37,500 square miles. Gold, as well as copper, exists in abundance throughout this vast stretch of country, judging by all indications." Bayles goes on to recommend the "Valdeze Pass" route which he claims his friend, Sunrise miner Billy Ribbstein used to explore the upper Copper River Route during the summer of 1897. He further adds ". . . the chances are strongly in favor of striking rich placer diggings before arriving at the head of navigation. I do not believe that a party starting on such a trip will have any occasion to push across the divide into the valley of the Yukon. You see the Klondike territory is pretty well covered by prospectors by this time, and by next spring the rush will be so great that the chances will be mighty slim for striking a good claim open to navigation (*The Alaskan*, 3/19/98)."

Whether these rumors were true or false, newspapers were soon reporting that "the Copper River country was richer than the Klondike, mining experts were buying up claims in the region and smelters and stamp mills would probably be erected at the mouth of the river by spring (Holeski, p. xix)."

Shipping companies, such as the Pacific Steam Whaling Company, immediately perceived a golden opportunity to capitalize on these press rumors. The company had operated the salmon cannery at Orca for the past ten years and was familiar with Prince William Sound. It was aware of the foolishness of schemes that promoted steamboat routes from the mouth of the Copper River. It was also familiar with the rumored Indian or Russian route over Valdez glacier and chose to promote that route instead. In what appears to have been a carefully orchestrated public relations campaign, the company set about to promote this yet unproven route.

In late September of 1897, Omar J. Humphrey, the company's Alaskan superintendent of canneries, and Supt. Brownelle of the Orca cannery arrived in Sitka where Humphrey issued a self-fulfilling prophecy that prospecting in the Prince William Sound and Copper River area was about to undergo a boom. He further predicted that the Valdes Glacier route was going to be to be the All-American to the Yukon (*The Alaskan*, 9/25/1897).

On October 11, 1897, Humphrey arrived in San Francisco with quartz samples allegedly assaying $20 a ton for gold and $10 a ton for copper. He boasted of a mountain on the shores of Prince William Sound composed almost entirely of quartz[!!!]. He further claimed that the head of the Copper River was rich in minerals — which is interesting since few non-Natives had ever visited the region. He further reiterated that the Valdez glacier trail would provide an easy, All-American route to the interior.

Following Humphrey's account in the *San Francisco Chronicle*, the company on October 21st announced that it would begin passenger service to Port Valdes and began its route on November 10, 1897 utilizing the retired revenue cutter *Wolcott*.

On the 18th of this month, Humphrey again appeared in the papers — this time on the front page of the *Seattle Post-Intelligencer*. Once more Humphrey promoted the Valdez route while decrying the route up the Copper River and the St. Michaels route. Humphrey hinted that unnamed telegraph and stage companies were already eyeing the Valdez route, and that his own company was considering regular steamer trips every 15 days from Seattle. Immediately after Humphrey's remarks, Pacific Steam purchased the steamers *Valencia*, *Excelsior* and *Alliance* for its Valdez passenger trade. (Holeski, p. xx-xxi).

One of the victims of this marketing hoax, William Treloar, from Carpenteria, California complains in his memoirs:

> We studied maps and charts and read the news papers and listened to the opinions of other people for a long time undecided which part of the country to go to until a representative of The Pacific Steam Whaling Company from San Francisco came down here and boosted the Copper River country. A country at that time greatly misrepresented. It is almost certain that no white man had ever been through that country at that time. But the San Francisco papers, that is the news papers especially the *Examiner*, told tales of wondrous rich strikes made there. And we were easily misled into choosing that country for a prospecting tour. Some great corporations are long experienced in to hoaxing the people and The Pacific Steam Whaling Company versed in that way of squeezing the Filthy lucre out of the public advertised that country and put all the boats they could spare on the route from San Francisco to Port Valdes via Seattle on purpose to catch suckers or anything that would bite (Treloar, Memoirs, pp. 1-2).

In a final act of journalistic irresponsibility, the *San Francisco Chronicle* on December 30, 1897 published a map of Alaska and the Yukon showing the locations of major gold fields. The same cross hatches that mark the Klondike and other proven gold fields indicate a large gold field on the west bank of the Copper River less than 80 miles northeast of Port Valdes. Readers did not bother to learn whether anybody had ever explored this region. Here was clear proof, here was something concrete, a map that proved gold was to be found on the upper Copper River. The stampede was on; thousands made the decision to go to Port Valdes, cross the glacier, and gather up the wealth lying on the banks of Copper River.

By April 1, 1898 *The Klondyke News* issued a somewhat belated warning to those who would dare the Valdez trail. Unfortunately, by the time of its publication, thousands were already ascending the glacier. Although the editorial is somewhat overstated and its readership would probably never have included those pouring over the glacier anyway, the balance it provides to the optimistic exaggerations of the San Francisco and Seattle papers is appropriate.

> We warn our readers against any attempt to reach the Klondike country by the way of Copper River. No living man ever made the trip, and the bones

of many a prospector whiten the way.

In the first place it is almost impossible to ascend the Copper River. There are trackless mountains to cross, by the side of which the Chilcoot Pass trail is a boulevard, and rapids that would make the White Horse dry up and quit business.

Certain unscrupulous parties operating steamboats up that way are issuing gaudy pamphlets with nicely worded directions of how to travel over a country that white man never set foot in. This is worse than murder, and such crimes deserve to be punished to the full extent of the law. We would suggest that they be hung, drawn, quartered and fed to a pack of hungry Malamute dogs (In Wharton, p. 211).

Despite occasional corrective warnings in the San Francisco and Seattle papers that the Copper River routes might not be feasible and that the mineral wealth of the Copper River was still unproven, rumor prevailed. Between 3500 and 4000 people from all across the country, indeed all round the world, made their way to Port Valdez in the early spring and summer of 1898. A number of the men and women who participated in this phase of the general "Klondike Gold Rush" left diaries, letters and memoirs describing their experiences on the Valdez Glacier Trail. In their writings we glimpse not only the motivations, the challenges, the hardships, the disappointments but also the joys and rewards of their perilous journey.

The Valdez Glacier Diarists

When a Chicago newspaper carrying the news of the rich Klondike strikes reached the small town of Freeport Illinois, French cook, Luther Guiteau, was just beginning to feel overtaken by middle age. The future seemed bleak and empty, and he yearned for some kind of adventure, some opportunity that would provide a real meaning to his life. He was surprised to discover that three of his contemporaries, Ed Kingsly, Bill Becker and Philo Snow, felt exactly the same way. As the four men discussed the Klondike news, they became caught up in their mutual enthusiasm. In early February, they found themselves disembarking a train in Seattle. Luther describes the scene:

The city was seething with activity; miners, would-be miners, stampeders, and sourdoughs filled the streets and bar-rooms and contributed to the steady flow of gossip and so-called news. One of the news items that affected our party especially was the report that "nuggets as big as bird's eggs" could be picked up on the banks of the Copper River.

Even with the considerable discount we prudently applied to the reports, the remainder was still glowing. We were at best a little late to get into the Klondike, we thought. After a conference, we decided to take passage for the foot of Valdez Glacier, on a little bay just off of Prince William Sound (Guiteau, *Golden Eggs* p.10).

13

Similarly, Joesph Bourke, the son of Irish immigrants in New York City had reached his 44th year. Artistically talented, Bourke had a decent job as an engraver for the U.S. mint. Still, the sense of malaise and restlessness that gripped the rest of the nation affected him also. Like the men from Freeport, he found himself caught up in the great rush to the Klondike. March 18, 1898 would find him arriving in Port Valdes aboard the overcrowded *Valencia*. Bourke left a perceptive diary of his experiences and became one of the founding fathers and mainstays of early Valdez.

Twenty-year-old Lillian Moore was one of the several women who joined the gold rush to Valdes. Vassar educated, single and independent, Lillian was a pioneer in more ways than one for she was in the vanguard of women testing their newly won freedom. Lillian joined a New York City expedition consisting of twenty women who pooled their resources to hire a schooner to sail them around the Horn and on to the goldfields of the "Klondike." Her talent with horses earned her the admiration of Captain Abercrombie who included her in his August expedition crossing Valdes Glacier. Her letter regarding this trip reveals that her motives were as much a search for adventure and opportunity than for gold.

In addition to a few single women like Lillian, a number of wives accompanied their husbands along the Valdes Glacier route often winning the admiration of the prospectors for their perseverance and hardihood. Some went on to prospect and file mining claims alongside their husbands while others started successful enterprises such as bakeries, restaurants, hotels and boarding houses. Lillian ran a successful livery stable before and after she met her husband Ed Wood.

Out of the north woods of Minnesota came Charles H. Remington and his brother, Grant. Under the pen name of "Copper River Joe," Charles later published a memoir recalling his experiences during the Valdez gold rush. His perceptions and judgments are more representative of those of the common man along the trail than the other writers who were mostly educated professionals. Joe's expressive, collo-quial narrative style catches the true flavor of story-telling of the day. Using his nickname to tell his story identifies him with his class. Addison Powell explains the use of nicknames in Valdez at this time.

> Like all frontier towns, many of the inhabitants of Valdez were known only by the "nick-names" which had become attached to them in some unknown manner. I was approached by a soldier who was enjoying a respite from Fort Liscum, and he inquired for the "Poor Man." I informed him that he was addressing the object of his search, but he refused to accept my view of the case . . . .
>
> An important day's doings at Valdez" that he had bought two dozen marten skins of "McKinley George" for a dollar each, but when he had attempted to sell them to "Cold and Greasy," he had been informed that they were muskrat skins, worth a nickel apiece; and that "Dad," "Alkali Ike," and

"Frenchy" had also declared them to be muskrat skins. . . . Those were familiar names in Valdez. (Powell, *Trailing and Camping in Alaska*, p. 182-183).

Like the others, thirty-four year old Horace Conger suddenly found himself stricken with Klondike fever. He abruptly sold his Minnesota pharmacy and set off to join the stampeders intending to cross the Valdes Glacier and prospect the rumored gold fields of the Copper River. His ambitions are revealed in a letter to his wife from Port Valdes on March 8th, "I hope and pray two years will find me on my way home with enough [wealth] to take us to Paris and to live on for the balance of our lives (Conger, Letter, 3/8/1898)."

Conger's willingness to risk everything on a single throw of the dice was typical of the period. One of the reasons that games of chance were so popular during the gold rush is that gambling was endemic to the whole enterprise. During this period, gamblers, if they were honest, were often referred to as merely "sporting men." When Conger failed to find gold in the Copper River, he traveled northward across Mentasta Pass to the headwaters of Seventy Mile Creek and down it to the Yukon River. He finally made his way home by means of the St. Michaels route. Thus, Conger was one of the few Valdes stampeders to have actually traveled this so-called All-American route to the Yukon gold fields.

Twenty-four year old Basil Austin was born and educated in England. Five years before the gold rush, he immigrated to the United States where he lived with his brother, John, in Detroit working as a machinist. The news of the Klondike had a tremendous appeal to this young immigrant just beginning to make his way in the new world. Recent immigrants and a number of foreigners, many of Swedish, Norwegian and Irish extraction, would make up a large percentage of those rushing to Port Valdes in 1898. Austin's decision to go to the Copper River was influenced by his meeting with a Mr. Demming of Detroit who was supposedly a local expert. Austin writes of him:

> He had accumulated quite a lot of knowledge about Alaska (true or false), and had some maps, particularly one of the Copper River valley that showed more than ours. He believed that all the good ground around the Klondyke would be taken long before we could get there, which was sound judgment, in short we should have to prospect in other territory. He believed the Copper River section was equally rich, more accessible and was by far the best proposition. Valdez, a new port near the mouth of the Copper, had been opened up, boats were running there and an easy pass over the Coast Range would bring us immediately to the best prospecting country of Alaska. (Austin, *The Diary of a '98er,* p. 2)

Of all the prospectors who crossed Valdes Glacier in '98, Austin was one of the very few who actually discovered gold in paying quantities, but it was not in the

Copper River area. After giving up on the Copper River country, Austin decided to go north to the Forty Mile mining district near Eagle. His route took him across Mentasta Pass and the Tanana River, past Lake Mansfield to Mosquito Creek, and down it to the Forty Mile. Here, he struck gold.

Diarist, Charles Margeson of Hornellseville, New York and photographer and diarist, Neal Benedict, of Seville, Florida joined a stock company expedition out of Stamford, Connecticut called "The Connecticut and Alaska Mining and Trading Association." (see the above cited *Prospectus*). This company, composed of 35 stockholders, most of whom were established business or professional men, sold stock to expedition members to finance a prospecting expedition to the Copper River region. By pooling their resources in a cooperative effort, they would be able not only to stake more claims but could afford more supplies than individual prospectors. Part of these supplies, they intended to sell in a company store.

These sharp Yankee traders had an early premonition that the real money to be made in the gold rush was from the rushers themselves. However, in supplying their own expedition their enterprising Yankee ingenuity failed them miserably; they brought with them an expensive and thoroughly impractical steam-sled with which they intended to cross the Valdez Glacier. Addison Powell muses retrospectively describing its debut on the glacier:

> Connecticut furnished a visionary company made up of persons who were distinguished from the others by having brought a steam-sled. All they wanted was to have the right direction pointed out to them, and they would steam over the glacier, ascend the Copper River, and stampede Indians, white men and every other thing encountered. . . .
>
> When the machine was steamed up and properly directed, the owners looked at each other disappointedly, for it failed to move. They applied the full limit of steam and it stood still some more, while the joke began to settle on Connecticut. The citizens should preserve that steam sled from vandalism as an evidence of the rushers of 1898. It had the record of being the "first automobile" in Alaska and was never guilty of exceeding the speed limit. (Powell, p. 29).

Thirty-five year old George Cheever Hazelet and forty-six year old Andrew Jackson Meals, two Nebraska family men, became caught up in the gold fever. Both hoped that by discovering gold on the Klondike, they would be able to benefit their family's economic status. Hazelet was a high school principal while Meals was a professional bullwhacker and frontiersman. His many practical skills would be of great benefit to their party on the trail. Both men and their families would play important roles in the development of Valdez and Cordova. On reaching Seattle, they decided to go to the Copper River rather than the Klondike. Hazelet's diary entry for

February 26 reads, "Have about decided to go to the Copper River and find we will have a hard time to get passage have been looking around all day. . . . Big stories in from Copper River and but few people there yet, so we will stand a better show than where there are so many."

Young Dr. Leroy Townsend of Beaver Falls, Pennsylvania was only 27 when he decided to join the Keystone Company organized by Col. H. R. Creighton of California. The Keystone Company like the Connecticut Company intended to engage in economic enterprises beyond prospecting. In addition to the search for gold claims, they built a store at Valdez and staked a townsite. Dr. Townsend was one of 25 doctors traveling the Valdez trail in '98. While many like Townsend were competent physicians, George Hazelet, contemptuously writes: "Some few quack doctors are located all along the trail, notable[sic] Dr. Matthews, Dr. Quick, Dr. Slow etc. If I had a sick dog I wanted to get rid of I would call them in but not otherwise." (Hazelet, 3/29/1898) Dr. Townsend was not one of these. Having enjoyed superior medical training at Philadelphia's Polytechnic University, he was called upon by other physicians in treating the scurvy epidemic at Copper Center during the winter of 1898-99. He played a heroic role in the treatment and late winter evacuation of some of the victims. Abercrombie, now a Captain, asked him to write a government report on the dread disease. Like the diarists cited above, letter-writer Townsend's immediate goal was to prospect the Copper River Country. Townsend wrote home from Seattle on February 3, 1898:

> All you see in the windows is Klondike outfitting & On the street, for advertising purposes, are Klondike sleds drawn by dogs of all sizes and colors. Various other Klondike machinery is being exhibited here and there.
>
> Klondikers are everywhere and are arriving at the rate of 1 or 2 hundred daily if reports here are true, and I have no reason to question; for on the train with us there were no less than 100. Most of these, too, strange to say were bound for Copper River. We are going to have lots of company. The reports are all favorable. Yesterday, though, a miner returned from Port Valdes and reported the pass still closed.

It is worthy of note that without exception these writers selected the route over Valdes Glacier not as an All-American route to the Klondike, but as a route to the supposed riches of the Copper River area. Certainly, shipping companies advertised it as an All-American Route and the military's mission was to establish such a route, but in the minds of the prospectors who came to Port Valdes in 1898, the goal was clearly the upper Copper River region. Thus, it is of little surprise that only a handful ever made it to the Yukon River over this route. Most were not headed there in the first place.

Chapter 2
The Gold Rushers
Arrive at Port Valdes

Early Arrivals at Orca

Just nine weeks after the nation's newspapers announced the depression-breaking news of prolific gold strikes in the Klondike, Adam Swan and thirty-three others sailed from San Francisco for Prince William Sound's Copper River. The party's leader, a wealthy California real-estate promoter, claimed that he had discovered gold in the Copper River Valley while serving on a U. S. revenue cutter. He confidently assured the company that small boats could ascend the Copper River passing through the coastal mountains to the interior.

In late October of 1897, the *La Ninfa* arrived at Orca, a small cannery town, now encompassed by Cordova. The disappointed prospectors learned from cannery personnel that no boats, large or small, could ascend the Copper River; it was impassable except when frozen. However, Chief Nicolais' band at Taral had brought copper nuggets from the upper river to trade at Nuchek.

Hearing this, some of the early gold seekers decided to remain near Orca waiting for freeze-up when they planned to ascend the Copper River from its mouth. They settled in at the nearby native village of Eyak "building sleds, arranging provisions into suitable packages for easy handling, and getting camp outfits into shape for use (C. H. Hubbard, *Diary*, p. 1)."

Around the cabin stoves at Orca, cannery personnel and perhaps other early arrivals, such as W. S. Amy, told Swan tales of another route across Valdez Glacier. Abercrombie had not been the only person to explore the glacier route. Amy and his party had arrived at Orca in September; they first tried the Copper River route, then chartered a boat to Port Valdez where exploratory hikes up the Lowe River and onto the glacier convinced them that the glacier was the better route. Orca resident, Jack Shephard, would have told of his and Pete Jackson's crossing of the glacier to Lake Klutina in 1896. And, of course, fur trader, Tom Olsen, had built the first cabin there. Olsen and Bill Beyer of Fox Island used to trade with the Ahtna Indians who brought their furs across the glacier. In fact, they would say that is how Beyer met his wife, Omelia. On one of the Indian trading trips, she had badly frozen her feet. The other Indians, unwilling to risk the lives of the rest of the party by attempting to carry her back across the glacier, had abandoned her to freeze or starve. Beyer discovered Omelia and nursed her back to health.

Early Prospecting in Prince William Sound

Prior to 1898, Prince William Sound prospectors fell into three general categories. First, the permanent residents — like Axel Lind, Tom Olsen, Bald-headed Chris Pedersen, Nils Jacobsen, Pete Jackson, Louis Thorstensen, W. J. Busby and Bill Beyer pursued a living by various means including fur trading, fox farming, fishing and prospecting. Second, summertime miners from the nearby Sunrise District prospected in the Sound; these included M. O. Gladhaugh, the discoverer of the Bonanza Mine at Virgin Bay, and William Ripstein, who also crossed Valdez Glacier into the Copper River district in 1897. Finally, "gentlemen prospectors" who rode the cannery boats from San Francisco, Sitka and Juneau staked a few claims and began planning the development of the Sound's mineral resources — men like U.S. Deputy Marshall for Alaska, William H. McNair, Deputy Clerk, Walton H. McNair, and Sitka grocery store owner, Ed de Groff.

Following the discovery of gold on Middleton Island on November 11th 1892, the Prince William Sound Mining District was established in 1893, rules for locating and recording a claim were set, and W. J. Busby was elected Recorder. Subsequently, quartz lode claims were located at Galena Bay in 1895 and 1896. The previous summer, Swedish immigrant, Tom Olsen, an agent for the Northern Trading Com—pany, Bald-Headed Chris Pedersen, and M. O. Gladhaugh, had found a rich, copper vein at Virgin Bay near the native village of Tatitlek and were now trying to develop it. The summer of 1897 had been a busy one for prospecting in the Sound. Beatson, Axel Lind and Olaf Carlsen had staked copper-bearing quartz lodes at Latouche Island, while a group from Juneau and Sitka staked placer claims in King Solomon's Basin at the head of Port Valdes and copper bearing quartz lodes at Landlocked Bay. However, unlike the placer deposits in the Klondike, which a sole prospector could mine, gold in the Sound occurred in bedrock; like the copper deposits, considerable capital would be required to develop the claims. Gladhaugh, Jacobsen, Thorstensen, Ripstein and others were looking for investors.

The First Stampeders arrive in Port Valdes (more on the spelling later)

On November 10, 1897, Adam Swan and about twenty others arrived in Port Valdes aboard the hired cannery tender, *Salmo*. They were put ashore behind a protected point on the southern side of the bay subsequently named "Swanport." Here, Swan and his party laid out the first townsite in Port Valdes. The party soon discovered that although Swanport was the best place to unload a ship, it was not the best location for starting the ascent of Valdes Glacier, since it was over four miles by water from the Valdes Glacier side of the bay. In addition, Swanport received no direct winter sunlight, while the southern facing Valdes Glacier side of the bay did.

Swan and his party sledded their gear across the ice at the head of the bay to the northeastern corner of Port Valdes near the mouth of Glacier Creek where the Pacific Steam Whaling Co. had already claimed 60 acres for a townsite they named "Copper

City." As the December days grew shorter and darker and the low, winter sun played peek-a-boo behind the peaks, Swan accepted a position as its Copper City agent and mail clerk and helped Capt. Zim Moore and Jack Shepherd build a cabin for the company's headquarters. While working together, the party must have learned more from Shepherd about the route over Valdes Glacier into the Copper River Valley and that now was not the time of year to cross it. It was better to wait until heavy winter snows had filled the crevasses, and the days grew longer.

The intense media campaign in the Seattle and San Francisco papers advertising the Copper River area as a prime gold area and the Valdes Glacier trail as a feasible route was having its effect. Two other boats, the *Wolcott* and *Behring Sea*, brought gold seekers to Port Valdes. Among the early arrivals was a party of twelve from Massachusetts. They settled two miles to the east of the Pacific Steam Whaling Company at the site of Tom Olsen and Bill Beyer's old trading post. On Christmas Day 1897, forty-five people found themselves together at the head of Port Valdes clustered in tents and cabins at Swanport, Copper City, and the old trading post site.

Establishing "The Law and Order Trail"

It is mid-winter. The cold nights are long and confining; space in the tents and cabins cramped. Earlier, high expectations remain unfulfilled; the wait stretches out beyond human control. Tensions build between members of the Massachusetts' party and Doc Tanner. According to a front page article in the February 2, 1898 *San Francisco Examiner*, the Massachusetts party while in Seattle met and invited Doc Tanner from Lexington, Kentucky and N. Call from Worthington, Minnesota to join them. Haines and Hogue, the leaders, promised to purchase six months' supplies for each member of the party and to land them at Orca. When Tanner indicated he did not have the full $250 to buy in, the others wanting his practical prospecting and all around mountaineering experience paid the additional $70. In what became one of the tragic ironies of the trail, Call gave Tanner his 44-caliber Colt revolver.

Tension between the easterners and the swaggering Kentucky gunman may have originated in the confined quarters on shipboard, but trouble began on their arrival at Orca. The Copper River route was not as portrayed. More disturbing, it appeared that instead of six months' supplies, there was only three months for each person. Tanner "took it harder than the balance of the party," and demanded an item–ized account of their expenditures. Haines and Hogue refused. Tempers simmered.

The group moved from Orca to Port Valdes where they found six feet of snow on the ground. It seemed better to build a cabin than to spread their sleeping bags on the cold snow at night. The work was strenuous, digging down through six feet of snow to the ground, falling and dragging the trees, raising the walls, and then — the leaders wanted to cover the roof with heavy, wet, gooey clay which had to be dug and hauled from two miles away.

Tempers flared. Tanner already upset, now became "unruly." After dinner, on January 2nd, he left the cabin for his tent. Whereupon, N. Call, W. A. Lee and B. F. S. Haines discussed Tanner's disruptive behavior and decided "the best way to keep peace in the party was to give Tanner his share of the supplies and let him go on his own resources."

To Tanner who had returned and overheard the conversation, this seemed like banishment — a sure death sentence in the cold, Alaskan winter. Tanner armed himself and returned saying "Well, I'm here for business. I heard all that you have said, . . . No man can bring me up here and throw me out in the cold." Whereupon, he quickly shot at Lee, Call and Haines. Lee and Call died within minutes. His shot missed Haines, the man he most wanted to kill.

Prospectors, transient residents of the unsettled frontier and boom towns, followed two codes of conduct: on the one hand, the frontier was a place to "live and let live." One minded one's own business and let others mind theirs, did not ask questions about a man's origins, and used nicknames instead of real ones. On the other hand, life on the trail and in the camps was perceived as the last opportunity for "true democracy" — equal participation by all on the local level.

Unlike Canadians who accepted the central authority of the Canadian government represented by the Mounted Police, U. S. citizens resented any intrusion by the federal government or military in civil affairs. Instead, the prospectors held miners' meetings to set rules and regulations. Decisions were by majority rule. Miners' courts, called in times of crisis, maintained law and order. In miners' courts a prospector had a chance to be a judge, jury member and policeman without the interference of attorneys and politicians. Unless the miners' decision conflicted with federal or state law, the courts would uphold them. At this time, of course, very few federal laws had been explicitly extended to Alaska.

In the Doc Tanner case, those present immediately summoned the other miners. The trial began at 11 o'clock that night. The miners elected a judge and were themselves the jury and executioners. Everyone had the right to participate and vote. Thirty-eight did, five refused. The one U.S. Marshall for all of Alaska was stationed in Sitka. Congress had not provided him any means of transportation to outlying areas or funds to transport a murderer to Sitka. This left the miners's court with few options: they could let Tanner go free, sentence him to a lashing or hang him. Since there was no jail or jailer and no one knew when the next boat would arrive, they did not have the option of shipping him to Sitka. Five hours later at 4 o'clock in the morning, the court reached its verdict: death by hanging at sunrise.

At 9 am on January 3, 1899, Tanner was hung from a tree near Olsen and Beyer's former trading center two miles east of Copper City. The site, henceforth, became known as "Hangtown" although later Valdez Townsite Trustees and the Chamber of Commerce preferred to call it "Old Town."

News of the hanging soon reached the outside papers. The February 2nd headlines for the *San Francisco Examiner* read in one inch type —

LYNCHED FOR TAKING TWO LIVES

'Doc' Tanner's Punishment at the Hands of Miners — Tried, Convicted and Executed at a Copper River Camp for Avenging an Alleged Wrong.

The editor prefaced the article with the statement — "Out of the rush to the north has come one more of the tales of horror to which men become accustomed when the lure of gold draws the reckless and unscrupulous in crowds from the restraints of civilization." Quickly, the sensational news spread across the country from San Francisco to New York. Not a miner going to Alaska, and especially those going to the Copper River area, missed the point. The Valdez trail was to be a "Law and Order" trail.

The Rush Begins —

As if the murders and hanging had broken the darkness, the days grew longer. On January 17th, a party of four men — Greenig, Bemis, Swanitz and Whitey Fish — carrying only packs crossed Valdez Glacier to Klutina Lake and returned. Although they nearly perished from the cold, they established the feasibility of the glacier route.

On January 16th and 28th, two more steamer-loads of eager stampeders arrived. And impatient prospectors began exploring the Port Valdez area. The Alaska Commercial Company, which operated the trading post at Nuchek, realized that a boom was in the making at Port Valdez and purchased land from Swan at Swanport hiring him as their agent for a competing trading post there. Simultaneously, the company, which had been primarily engaged in the fur trade, began negotiating with the old-timers for the development rights to copper claims at Virgin (Ellamar) and Landlocked Bays.

Winter storms covered the land with foot upon foot of snow, crevasses in the glacier filled, the sun rose higher clearing the mountains above Swanport, and the days grew ever longer. On February 14th, early arrivals, like W. S. Amy, began pulling their gear up the glacier closer and closer to the promised land. Looking back down the glacier, they could see more steamers arriving and more adventurous gold seekers coming behind them.

Spring 1898 — The Stampede

On arriving in Port Valdes in early March, George Hazelet complains laconically to his diary, "It was claimed when we left Seattle that there was a wharf at this point, but if ice 18 inches thick is a wharf then we unloaded on a wharf (Hazelet, 3/13/98)." However, compared to ships arriving later, Hazelet's party did indeed enjoy a unique "ice wharf." Gold seekers arriving in February and early March encountered a thick layer of ice covering the tidal flats at the head of the bay. For these early stampeders, the ice turned out to be more of a blessing than a curse; for in the absence of any kind of dock or wharf, the ice provided a stable platform close to deep water where steamers or schooners could unload their passengers and tons of supplies. The ships in late winter needed only edge up alongside the ice sheet at high water. Once the ship was made fast, a boom was extended over the side and the prospectors' goods were deposited onto the ice. Then the sorting and sledding would begin.

Once on the ice, it was an easy matter to sled the goods the mile and a half to the cottonwood groves on the shore sheltering a random collection of tents known first as "Copper City," then as "Camp Valdes," later as "Port Valdes," or simply "Valdes," and finally "Valdez."

As shipload after shipload of eager gold seekers arrived, the settlement grew so that by mid-March, one frame building and from 100-200 tents housing between 700-900 inhabitants lined the glacier stream. The rapidly growing little community possessed a dynamism all its own. "This unique camp — for it was about that — presented a scene of unusual activity. Some were tramping down the snow, preparing a place to set up their tents; some were cutting tent poles, and others cutting firewood, while others were getting their dog teams ready for hauling their goods up to the foot of the glacier, which was five miles away (Margeson, p. 55)."

"It was claimed when we left Seattle that there was a wharf at this point, but if ice 18 inches thick is a wharf then we unloaded on a wharf (Hazelet, 3/13/98)."

"Arriving at Valdez, we found no wharf nor storehouse or any other convenience, nothing in fact but a snow bank (Bourke, Journal, p. 4)."

Photograph by Neal Benedict, who was a member of the Connecticut and Alaska Mining and Trading Co. From the Messer Collection courtesy of the Cook Inlet Historical Society.

"This unique camp — for it was about that — presented a scene of unusual activity. Some were tramping down the snow, preparing a place to set up their tents; some were cutting tent poles, and others cutting firewood, while others were getting their dog teams ready for hauling their goods up to the foot of the glacier, which was five miles away (Margeson, p. 55)."

Photograph by Neal Benedict from the Messer Collection, Courtesy of the Cook Inlet Historical Society.

Treloar who arrived on the *Valencia* that windy March day remarks that the little tent town even offered some civilized amenities. He describes his first evening ashore eating out at a Valdes restaurant.

> Bill Miller and I had our first meal at the only restaurant there was at Port Valdes, only ten men could eat at the same time, there were ten men inside when we went there, and we had to wait until they had finished. The restaurant was a tent, set down in the snow about ten feet. The camp stove was the only thing on which they had to cook, so it took some time to get a meal. While we were waiting, we picked out a place to put our tent, we marked it and put up our sign, then we went back to the restaurant. We sat down a few minutes until the landlady announced dinner, she told us to come right in, and make ourselves at home, but it didn't look very homelike, the tent was about eight by ten, a box for a table, and a few more for chairs, the dishes were granite ware, and beans, bacon, baking-powder bread, cooked dried potatoes, and coffee was our bill of fare. We were glad to get even this bill of fare for we had been on short rations on the boat for three or four days, sour bread and salt horse, so this tasted pretty good to us. We filled up on it. Before we got through it made us think of home sure enough we paid one dollar for our meal ... (Treloar, Memoirs. 15-16).

However, the amenities offered by his tent pitched on the bare snow that first night did not quite measure up to his experience at the town's restaurant:

> ... sleeping on the snow is not what is cracked up to be. I can only speak for myself but I believe the rest of the boys were in the same fix I was. First we would lay on one side and the underside of a fellow gets pretty cold lying close to the snow when it got so cold we could not stand or endure it any longer we would flop over to the other side. That was a long night for me and I was glad when morning came (Treloar, Memoirs, p. 17).

Toward the middle of March, the temperatures warmed and the ice wharf became more precarious. The first signs of trouble appeared on March 15th. As the Connecticut Company unloaded its steam-sled and heavy lumber, a sudden thud reverberated across the ice as it settled onto its mud base. Eight inches of salt water oozed across the breached surface. The men struggled to move their endangered goods and precious steam-sled to higher ground. While they hastened to unload the remaining 35 tons of groceries and hardware, the situation worsened.

> Toward noon, the third day, the ice had become so rotten that large cakes became detached, and would settle under us into the water as we passed over them with our goods. The last half day of this work was attended with great danger, for had any of us broken through with our loads, drowning would have been the almost certain result. I remember well one time, while drawing my load, that I stepped on a large cake, which broke into several smaller ones with my weight, none of which were large enough to support me, and leaving my load, I sprang lightly from cake to cake until I had reached firm ice; then, getting the assistance of some of my companions, we carried lumber and bridged the spot sufficiently to bring over the balance of our goods. We were all glad when — the last of it was got off; for there was scarcely a man but could relate some thrilling adventure during his work over the rotten ice. (Margeson, p. 57)

By the time of the arrival of the *Valencia* on that windy March day the fragmented ice had become detached from the shore and blown by the strong winds down the bay. The ice wharf could be depended on no more. The lack of a suitable place to unload off the glacier stream caused the Captain of the *Valencia* to select Swanport as a point of debarkation and precipitated the ensuing passenger revolt. The captain was forced to call in flat bottomed, lightering vessels from the company's cannery at Orca.

CAMP VALDES (Benedict):"This is an awful country. Nothing but snow and ice as far as the eye can see. Where we are now camped [on the beach] the snow is 7 ft. deep. Of course we do not dig down to the ground to pitch our tent. Will be here 3 days more and then will move our tents 5 miles farther up [to the foot of] the glacier." Conger, 3/13/98.

Photograph by Neal Benedict from the Messer Collection, courtesy of the Cook Inlet Historical Society.

While the disappearance of the ice wharf at the head of the bay was a headache for shipping companies, it provided a boom for some of the early residents of the Sound who almost overnight had become "old-timers." The early Orca cannery workers and local fox farmers, who were themselves part-time prospectors, must have looked upon this sudden invasion by several thousand greenhorn prospectors with a mixture of amazement, apprehension and disdain. However, a number of them came to Port Valdes to make a first-hand appraisal and see if perhaps a little extra cash might be made. For example, Louie Thorstensen brought his sloop in from his fox farm on Goose Island and employed it as a ferry and lightering vessel. He also chartered to geologist, Frank Schrader, for his early reconnaissance of the Sound. Cannery workers could earn 2 1/2 ¢ a pound packing prospector's goods across the mudflats from the lightering vessels. A man could make as much $12 to $15 a day in this manner —fantastic wages in those days. Fox farmers and prospectors Middleton Island Smith, Bill Beyer and Bill Busby would play an active role in the establishing of the Valdez townsite. Fur trader and cannery worker, Pete Jackson, gave up his job at the Orca Cannery to become the new mail carrier along the trail. W. E. Ripstein, one of the pre-98 prospectors, obtained power-of-attorney from his fellow prospectors and began selling their Prince William Sound claims.

The lightering operations involved considerably more effort than the earlier ice wharf. On April 18th, the military's Copper River Exploring Expedition under now Captain Abercrombie employed two small, flat-bottomed cannery steamers to unload their 20 tons of supplies off the *Valencia*. The loaded steamers were allowed to go dry on the mud flats at low tide. Then men in hip boots carried the supplies on their backs through the muck a quarter mile to the 6 foot snow scarp on shore. From here they hauled their loads one-half mile over deep snow to the little tent settlement. Captain Abercrombie describes the condition of his men the night after this ordeal:

> This was one of the most trying nights that the expedition was to experience during the explorations of 1898. The officers and men, not being used to the work they were called on to perform, were badly bruised about the shoulders and back. As the day was comparatively warm and the exercise violent, the clothing of the men became dampened, and in this condition they were forced to spread their blankets and sleeping bags on the snow by the cache without outer covering and thus pass the night. The thermometer registered 8° below zero, but not a grumble was heard from any of the party, although at daylight the next morning some of the men were beating their frozen boots on the crust of the snow so as to get them in condition to pull on their feet (Glenn & Abercrombie, p. 299).

Abercrombie here indulges in his characteristic exaggeration when reporting to his superiors. Although 1898 was at the end of "The Little Ice Age" and tempera–tures were colder than now, 8° F. below zero was extremely unlikely for this late in the spring. Conger on the colder glacier summit reports 10° F. above for this date.

Chapter 3
The Founding of Valdez

The Merchants Arrive

Between December of 1897 and March of 1898, Valdes ceased to be just a landing place and became a town with merchants providing goods and services to the gold seekers. Just as unloading supplies posed more of a challenge, so building a store proved more difficult than anticipated. The Connecticut Co. discovered no warehouses on shore in which to store their merchandise, only a snow-covered beach exposed to the weather. They began shoveling down through the ten feet of snow, then towed it away by hand-sled. Trees had to be fallen, hewn, and carried by hand to the building site. It took fifteen men several days to finish the store.

When the store opened around March 20th, they found ready customers. After spending a number of uncomfortable nights sleeping on the bare snow and ice, George Hazelet writes "went down to the landing and brought some lumber of Con [Connecticut] and Alaska Trading and Mining Co. Got enough to cover the bottom of tent 100 ft. (*Hazelet Journal*, 3/29/98)." The hustling Connecticut Co. men even took the seventy-five mattress on which they had slept during the trip up and "quickly sold them for $3.50 each" at a 300 percent profit. The store was such a success that they sent Mr. Potts back to Seattle to purchase more goods — especially tents, tarps and lumber which were all in short supply. An item purchased in Seattle "brought two to four times its cost (Margeson, p. 66)."

COPPER CITY (Bourke):"This is going to be a boom camp, I tell you. Reports say that the Pacific Whaling Co. now have 5000 booked for this place - Valdes — and it is estimated that anywhere from 6-40 thousand will come in (Townsend, 3/30/98)."

Photograph from the Joseph Bourke Scrapbook courtesy of the City of Valdez.

The Connecticut Co.'s Smitthy: "Many of the men who preceded us had begun the ascent of the glacier, and there came a great demand for "ice-creepers." We had taken along a blacksmith's forge and tools, and a large quantity of sheet steel, and having two blacksmiths in our company, we set one to making "creepers" (Margeson)."

Photograph by Neal Benedict from the Messer Collection, Courtesy of the Cook Inlet Historical Society.

One item in especially short demand was ice-creepers which when fitted to one's boots provided a better grip on the ice. Townsend, writing from the Keystone Group's townsite of Port Valdes, says, "When we landed here, it was upon the ice and about a mile from where we are camped. I tell you, it was hard work to pack our goods in to land. The ice was smooth and rough by turns, and we had no ice-creepers. The Col. thought we would have no use for them (Townsend, Letter, 3/8/98)." The Connecticut Co. men knew a market when they saw one. They quickly set one of their two blacksmiths to work making ice creepers from their "large quantity of sheet steel." He could make a dozen pairs a day, which sold for $3.50 each. "Thus we were able to reap a large revenue off the labors of one man (Margeson, p. 66)."

Prospectors were not the only ones to believe the steamship company's hoax, so were many merchants who came to "mine the miners." Bryan Pearson reports that when he arrived on April 20th, Valdes had "four log cabins, three frame houses, and many tents. Three restaurants are in full swing, sundry stores and of course a saloon, the proprietor of which evades the strict import laws by calling his drinks 'stomach bitters' (Pearson, p. 185)." When Will Crary made a quick trip back to Valdez on June 11th for mail, he stayed at Gray's Hotel. There was even a photography shop where a prospector could have a photo taken of himself in Valdez to send home.

The Townsite Rush

In addition to locating claims for their stockholders, companies like the Keystone Co. and Connecticut Co, intended to profit by establishing townsites. An early *Prospectus* by Harry E. F. King for the Connecticut Co. outlines their plan:

If we find gold in large quantities we intend to stake out a townsite, build a block house to live in, put up our saw-mill, establish a store or trading station, dig out some gold and send it down the river by the crew of the schooner and start them back to Seattle, show the gold and make an hurrah, boom the

The Connecticut Co.'s store was an immediate commercial success. Dr. Kortright, one of the party's members who turned back on the glacier, played an important role in the founding of Valdez as a member of the second townsite committee. Several townsite committee meetings were held in the store. Note the two children in the photograph.

Photograph from J. Bourke Scrap-book, courtesy of the City of Valdez.

location, load the schooner with provisions and a number of men to work the mines.

Crowds of prospectors will rush in then we can sell them claims, building lots, provisions, lumber, tools or anything they want to buy, including medicine and then the fortune we are going to look for will be [*sic*] simply float in without much severe work on the part of the company. . . . The officers of this Association are competent mechanics and receive no salary for their services, every one a hustler. (*Prospectus of the Connecticut and Alaska Mining and Trading Association,* Crary Scrapbook, vol. 1, p. 112).

Between November 1987 and mid-March 1898, would-be developers claimed at least seven townsites at the head of Port Valdes, some of them overlapping:

1) Swanport on the southeastern side of Port Valdes (Nov. 1897)
2) Pacific Steam Whaling Co.'s 60 acres for Copper City on the northeast side of Port Valdes on the route to Valdes Glacier, (Dec. 1897);
3) Adam Swan's 1/4 by 1/4 mile townsite adjacent to Copper City (Dec. 1897);
4) Hangtown, a little less than two miles to the east of Copper City;
5) The Portland Groups' townsite adjacent to the Pacific Steam Whaling Co. (March 4, 1898).
6) The Keystone Company's site, "Port Valdes," adjacent to the Pacific Steam Whaling Co. (March 5, 1898).
7) The J. M. Hayse townsite, east side of Glacier Creek towards Corbin & Valdes Glaciers (prior to March 6, 1898).

Simultaneously, two other groups claimed use of the land: the merchants. and the gold rushers, who camped for about a week in Valdes while moving their supplies to the foot of the Glacier.

And, the population was growing rapidly. No longer did everyone know everyone else. From the 45 prospectors who spent Christmas 1897 together, the population grew to 600 by early March and from 1200 to 1500 by the end of March.

When Pearson arrived on April 20th, he reported 2000 people in Valdez. In his March 30th letter, Dr. Townsend of the Keystone Co. writes:

> This is going to be a boom camp, I tell you. Reports say that the Pacific Whaling Co. now have 5000 booked for this place — Valdes — and it is estimated that anywhere from 6-40 thousand will come in. I place my estimate in the smaller figures. Boats are coming in every few days. They are mostly of small tonnage though, and 20-60 passengers (Townsend, 3/30/98).

The growing number of townsites pitted two groups against one another. On the one side, gold rush entrepreneurs wanted to acquire free public land for real estate developments. On the other side, citizens and merchants demanded free lots.

Merchants wanted the townsite so they could obtain title to lots and protect business investments. But this was not the only reason. As the winter snow melted, the signs of three to four thousand people camping without outhouses became appalling apparent. Residents were concerned about sanitation and the safety of the community's drinking water. The large number of transient prospectors, many from urban backgrounds, were unfamiliar with guns, resulting in accidents from the misuse of firearms. Miners' and citizens' meetings could handle crisis situations, but citizens felt the need for a stable local government which could pass and enforce ordinances protecting public health and safety.

The pressure to establish a legally recognized Valdez townsite gained even more momentum when on April 21st Capt. Abercrombie suddenly claimed most of the area staked by the Pacific Steam Whaling Company as a military reservation. With a stroke of the military pen, Copper City ceased to exist. It was a warning that business entrepreneurs and citizens could easily lose their investments.

According to Bourke, "the first survey of the town of Valdez was begun April 3, 1898 (Bourke, Notes, p. 247)." Citizens met to form a townsite on April 5th. On April 21st, the boundary line between the military reservation and townsite was surveyed. The townsite area selected was staked over the claims of the Keystone Co. and the Portland group.

On April 23rd, following the familiar model of the miners' meetings, "a mass meeting of citizens assembled to organize a town or city (Minutes, Crary Scrapbook, p. 76a)." The citizens' meeting included old-timers, soldiers, businessmen belonging to companies like Connecticut Co., mom & pop shop owners and prospectors. Old-timer W. C. L. Beyer was elected chairman of the citizens' meeting.

The townsite plat, which located streets, blocks and lots, became the subject of considerable controversy. The citizens wanted to review it; the township committee refused. Consequently, over a period of several weeks, the citizens elected a township committee to serve one year terms, then disbanded the township committee and voted in a townsite committee whose members held "office till discharged by the citizens (Minutes, 5/10/1898)." And, adhering to a fiercely democratic frontier

This may have been the building Adam Swan helped to construct for the Pacific Steam Whaling Company. A photograph taken later in the summer has a sign posted by the entrance way indicating that this is the ticket office for the Pacific Steam Whaling Company. In 1899, the Pacific Steam Whaling Co. Store was converted into a hotel.

Photograph from the Bourke Scrapbook courtesy of the City of Valdez.

spirit, they immediately passed motions concerning the public notifications of meetings: "meetings must be posted 24 hours" and five notices must be posted in public places (Minutes, 5/10/1898). Today, almost a century later, the 24 hour requirement remains in effect, but the number of postings has been reduced to three.

On May 13, Frank Kertchem, a member of the Portland Group whose townsite claim had been incorporated into the new townsite, wrote to the Honorable B. Hermann in the General Land Office in Washington, D. C. inquiring about the law:

> I with several others from Portland Oregon landed here on the third day of March — this year — on the fifth, we posted a notice stating that we had taken up 160 acres of land for trade purposes under the act of March 1891. . . .
>
> In the last two weeks some parties have jumped the claims of the "Keystone" company (a trader's claim) and have organized a town com.[committee] and have surveyed same in part and they have voted our claim into the town limits — against our will. Can they do this legally? (Kertchem, Letter, 5/13/98).

Questions over the legality of the Valdez Townsite delayed acquisition of the townsite patent for over a decade until U.S. Patent No. 273637 was issued to W. H. Crary, Trustee, on June 11, 1912. Ultimately, the founding of Valdez illustrated the old frontier saying that "Possession is ninety-nine percent of the law."

On May 14, 1898, the citizens finally adopted a townsite plat. Unlike the helter-skelter plats of many mining communities, the Valdez plat shows a city based on the principles order and consistency; a townsite designed to bring order to the chaos of the wilderness and consistency to the lives of the people who settled there. Ironically, during the entire debate, no one apparently asked whether the townsite was located in an topographically suitable place, one protected from severe glacier winds, seasonal flooding, or unstable sediments. Edward Gillette, railroad engineer on the Copper River Exploring Expedition, expressed his concerns in his Report:

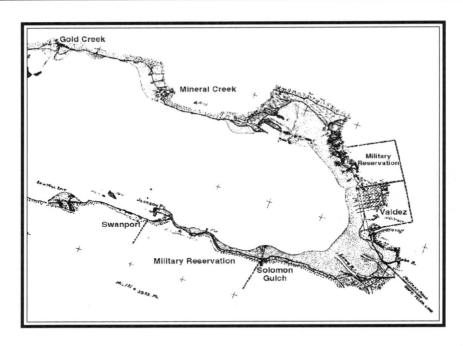

In November 1897, Swan located the first townsite at Swanport. On April 21, 1898, the military made its first reservation on the NE side of Port Valdez. On April 23rd, the citizens of Valdez claimed a 1/4 square mile townsite. This was later expanded to a square mile. In 1899, Abercrombie selected a second military reservation on the SE side of Port Valdez. This site later became Dayville, then the Alyeska Marine Terminal. Following massive subsidence of unstable soils in the 1964 earthquake in 1964, the City of Valdez was relocated to the Hazelet townsite by Mineral Creek. (Map courtesy of the Valdez Historical Society).

... where the town of Valdez has been hastily built there is danger at any time of having the buildings swept into the bay by the swift and quickly changing channels formed by the numerous streams flowing from uncertain and ever changing parts of the immense Valdez Glacier, situated some 4 miles north of the town. An occurrence of this nature would doubtless cause the loss of many lives." (Gillette, Report, Abercrombie, 1900, p. 138-139).

In 1899, 1900, and 1901 the Valdez Glacier stream severely flooded the town. While merchants and townspeople raised subscriptions and later obtained federal funding to build dikes to protect against flooding, Abercrombie, in 1899, decided to relocate the military reservation to higher ground on the other side of the bay. His decision to abandon the Valdez site was probably also influenced by Gillett's identifying land adjacent to the new military reservation area as the best place for a railroad terminus.

An examination of the city's minutes shows the citizens retained the right to approve ordinances, make financial decisions, and dispense justice. The townsite committee generally served more of an advisory role than a legislative one. In mass assemblies, the citizens immediately passed ordinances giving women the right to own lots, requiring sanitation and garbage disposal, prohibiting shooting within 800 yards of the townsite, and protecting the town's drinking water. They changed the Townsite Committee's proposed voter requirement from thirty days residency to two weeks and refused to pass an ordinance setting a uniform punishment for those convicted of violations.

Old-timers: W. J. Busby

Two men stand out in the early days of Valdez: old-timer, W. J. Busby and newcomer, Melvin Dempsey. Busby, a fox farmer who filed his first claim in Prince William Sound in 1893, was the Recorder for the Prince William Sound Mining District. Prior to the Valdez gold rush, Busby was the second most important official in Prince William Sound after the customs inspector at Orca. On April 25th, Busby was elected to the first Valdez Township Committee and elected Recorder for both the old and new township committees. During the month of May, he recorded 95 townsite lot claims, but then suddenly returned to his fox farm on Busby Island. In June, July, and August only one lot a month is recorded. At a citizens' meeting on October 10th, the office of the Valdez Recorder was declared vacant.

During April and May, Busby was embroiled in three major controversies. The first involved the citizens' township committee's jumping of the Portland Group's townsite claim. The second, as mentioned above, occurred over the townsite plat where Busby was caught between the irate citizens with whom he apparently sympathized demanding the town plat and the townsite committee on which he served refusing to give it to them. A third controversy erupted when the Portland Group, Dempsey and others created a separate Valdez Mining District. Again, we have a record of this from Kertchem's letter to Hermann:

The Valdez townsite was close to the Valdez Glacier Trail, but it was also located on the glacier's outwash plain and subject to severe flooding. In 1899, 1900, and 1901 the town experienced frequent flooding. Numerous dikes were built to protect the town. Capt. W.R. Abercrombie relocated the military reservation to Ft. Liscum to avoid the flooding.

Photograph No. B62.1.2129 from the Crary Collection, courtesy of The Anchorage Museum of History & Art.

Some gold rushers came to "mine the miners." Men like J. J. Riggins and "Oklahoma Bill" Hemple purchased returning prospector's outfits then resold them. J.J. Riggins replaced old-timer, W. J. Busby, as Valdez's townsite Recorder in October 1898. In January 1899, when the Valdez Mining District was officially formed, J.J. Riggins became the Recorder and his brother Chas. Riggins the deputy recorder.

Photograph from Joseph Bourke Scrapbook, Courtesy of the City of Valdez.

Some parties since our occupancy of said claim, state that we could not form a new mining district without first having notified the recorder of next adjoining district some 28 miles distant — nor having posted notices for 10 days prior to said meeting — now I understand the mining laws of Oregon apply where nothing special for Alaska exists. Is this so? (Kertchem Letter, 5/13/1898).

The conflict over mining districts simmered throughout the summer and fall as the newly arrived prospectors and merchants, who decided not to go over the glacier, began staking claims in Port Valdez. Between mid-March 1898 and January 1899, over thirty claims were recorded; some of these were in the Solomon Gulch area where earlier prospectors had already staked claims in 1897. Other new-timers, including H.E. Fenske staked quartz lode claims at Glacier Island. Most of those in Valdez had no way of contacting Busby at his home on Fox Island near Tatitlek. By mid-winter 1899, it was evident to the Valdez prospectors that to assure the legality of their claims Busby's criticisms that they had illegally formed a new mining district had to be met. Dempsey and others called a meeting at the Endeavor Hall giving the old-timers Busby and Jack Shepherd, at Orca, a written thirty days notice that they were establishing a new mining district. On January 16th 1899, the Port Valdez Mining District was formed. All claims within Port Valdez would be filed with the new district, those outside the Port with the Prince William Sound Mining District. Townsite Committee Trustee, J.J. Riggins, who replaced Busby as the city's Recorder became the Recorder for the Valdez Mining District. Busby remained in the Sound. He had had enough of Valdez politics.

Other old-timers apparently agreed. They remained at their homes in the Sound and continued to prospect. Of the thirty prospectors who made claims in both 1897 and 1898, none of the 1897 prospectors caught the Copper River gold rush fever.

They continued prospecting around the Sound's known mineralized areas — Copper Mountain, Landlocked, Galena and Virgin Bays, Latouche and Knight Islands.

The gold rush to Valdez provided the old-timers a market for their mineral claims in the Sound. Some prosectors began selling their properties to each other and to the investors and developers who followed the gold rushers. Others combined their knowledge of the Sound's mineral resources with their skills as salesmen to broker claims for their fellow old-timers. Old-timers from Sunrise, W.E. Hunt and S.J. Mills (the Recorder for the Sunrise District), who had prospected in the Sound in 1897, began purchasing. In May, they purchased one of M.O. Gladhaugh's claims at Virgin Bay. F.C. Lawrence from Ashcroft, British Columbia, one of the "gentlemen prospectors" who visited the Sound in 1897 purchased another claim from Gladhaugh for $40,500. W.E. Hunt and William Ripstein also brokered claims for their fellow prospectors. In September 1898, Ripstein sold Phil Blumauer, a newcomer who was a member of the first township committee, seven of the old-time prospector's claims on Latouche Island. A year later, in a rare joint venture between old-timers and new-comers, early prospector Chris Christiansen joined with Blumauer, Capt. Abercrombie and others to locate and record claims on the east and west sides of Port Wells. However, W.E. Hunt made the most interesting sale. In 1898, for $10,000 he brokered four copper claims on Latouche Island to Capt. Omar J. Humphrey, the man largely responsible for booming the Valdez Glacier Trail. Humphrey, himself, did not invest in early Valdez. He placed his money on the proven earlier claims in Prince William Sound.

New-comers: Melvin Dempsey

Melvin Dempsey, one of the newcomers, arrived aboard the *Alliance* on February 26th 1898. He was both a merchant and a prospector. On March 23rd, he opened his restaurant; on April 24th, he organized the Valdez branch of the Christian Endeavor Society; on May 9th, he called the citizens' meeting that led to the ouster of the first township committee; on May 12th, the new townsite committee presented the Endeavor Society with a free lot for their Hall and Reading Room. From May 31st to July 9th, he traveled over the Glacier to the Copper River and back; on July 23rd, he was appointed to draft by-laws for the Valdez Mining District (Book No. 1 of Valdez Mining Record). On September 6th, he reports the arrival of the U.S. steamer *Wheeling* with Alaska's Governor John Brady on board. On September 18th, he attended a meeting for a relief station to which the town treasury donated $10.

Melvin Dempsey is most remembered for his efforts to help prospectors. As often happens, those that benefited most financially from an event, in this case both the shipping companies and merchants in Seattle and San Francisco, were not the ones to bear its financial and moral burdens. These fell to the merchants and citizens of Valdez and the federal government. We may judge the moral fiber of a community

Melvin Dempsey (standing in the doorway) organized the Valdez branch of the Christian Endeavor Society on April 24, 1898. He was instrumental in constructing and providing for both relief stations on the glacier and the Endeavor Society Hall and Reading Room. During the winter of 1898-1899, the Hall served as the civic and cultural center of Valdez. Townsite committee meetings, inquests, religious and social functions all occurred here. Photograph No. B62.1.951 from the Crary Collection, courtesy of The Anchorage Museum of History & Art.

by its care for its transients. Dempsey's name stands out among the early citizens of Valdez for his concern about the welfare of prospectors crossing the glacier.

In September, Dempsey anticipating the hardships of miners returning over the glacier led Endeavor Society members in soliciting funds and organizing the construction and maintenance of a 10 by 12 foot relief hut on the glacier supplied with food, fuel, and bedding. He also appealed to the town's merchants and military for contributions. The Pacific Steam Whaling Co. contributed lumber for the building and 58 gallons of coal oil which indigent prospectors hauled to the site in exchange for their meals; the Keystone Company and Connecticut Co. contributed money to the relief station's construction and subscribed to a relief fund for provisioning it. Although over fifty people, including five women, contributed, Dempsey's name is most frequently associated with the relief station. On October 7th 1898 its Endeavor Hall and Reading Room, a 14 by 20 foot log house, that became the young town's civic, social and intellectual center. According to Bourke, Dempsey "contributed continuous labor and supplies for which no charge was made." He also credits Dempsey with starting "the C. E. [Christian Endeavor] tent on the beach, both relief station plans, the church and the log reading room (Bourke, CS, p. 247)."

On October 4th, Dempsey records in his diary simply, "I was elected one of town trustees." For a brief period in his life, the opportunities of the frontier, which was still free from the petty conventions and pressures of civilized society, had opened up possibilities for Dempsey which otherwise were closed to him. He was the descendant of a Cherokee Indian plantation owner and a Negro slave.

How do you pronounce the Town's Name?

One important item that had to be settled by the townspeople was the exact spelling and pronunciation of the name of the new town. Lt. Salvador Fidalgo in surveying the area for the Spanish crown in 1791 named the bay forming the northeast corner of the Sound "Puerto Valdes." Vancouver later anglicized this to "Port Valdes." Prospectors in 1898 were told their mailing address would be c/o The Pacific Steam Whaling Company, Port Valdes, Alaska. Like most people today, they would have naturally pronounced the name as the Spanish do with a short "e" and a final "s." However, when they arrived, they discovered that old-timers like Beyer, Busby, and Hall had further Americanized the name pronouncing it with a long "e" and a final "z" as we do today. We know this because we have an annotated diary in the Valdez Heritage Center Archives by Horace Tuffin who accompanied George Hazelet on his second expedition (1900). Tuffin's diary, like many documents of the period, vacillates back and forth between spellings of "Valdes" and "Valdez." An annotator who accompanied Tuffin corrects one of these spellings writing in pen on the manuscript "Old-timers called it Valdees." Also, *The Seattle Times* interview cited above quotes old timer and trapper Bayles as referring to "Valdeze Pass." (*The Alaskan* 3/19/98). Hence, the spelling "Valdez" appears to be a deliberate attempt to capture the actual pronunciation of the old-timers.

The city founders soon became divided as to the town's name and how it should be spelled and pronounced. This division seems to have been between two groups — the old-timers (pre-'98) and the newcomers of '98. The earliest minutes of the township committee, April 23, 1898, are headed "Valdez" as are those of the second meeting on April 25. However, the second meeting's minutes record "This commit-tee after due consideration would submit to you that the town be called 'Port Valdez.'" Old-timer Busby was the secretary for these two meetings. However, on April 26th, newcomer, Dr. Kortright of Connecticut Co. takes over as secretary and the city becomes "Port Valdes." When the citizens gather in a mass meeting on May 19th the Secretary becomes old-timer Hall and "Port" is dropped from the town's name and the final "s" is changed back to "z." From this time forward the town is always referred to as "Valdez" or "Valdes" depending upon which group is recording the minutes. That the pronunciation and spelling of the name remained contentious is evidenced in the July 18th minutes of the townsite commissioners meeting which records "moved the official name of this town be called Valdes, spelled VALDES" [emphasis in original]. However, since Hall continues to record the minutes, the spelling remains "Valdez." On August 5, newcomer John Snyder becomes secretary, and the name reverts back to "Valdes" until April 3 of 1899 when it inexplicably is changed back to "Valdez." Finally, on May 13, 1899 the town was granted a U.S. Postal Station and "Valdez" became official.

An oral tradition still current in the town maintains that the Americanized spelling won out in the end because of bitterness toward the Spanish during the Spanish-America war which was fought during this period. Even today, newcomers still insist on pronouncing the name of the town as if it were Spanish ending in an "s," while we old-timers pride ourselves in the correct pronunciation "Valdeez."

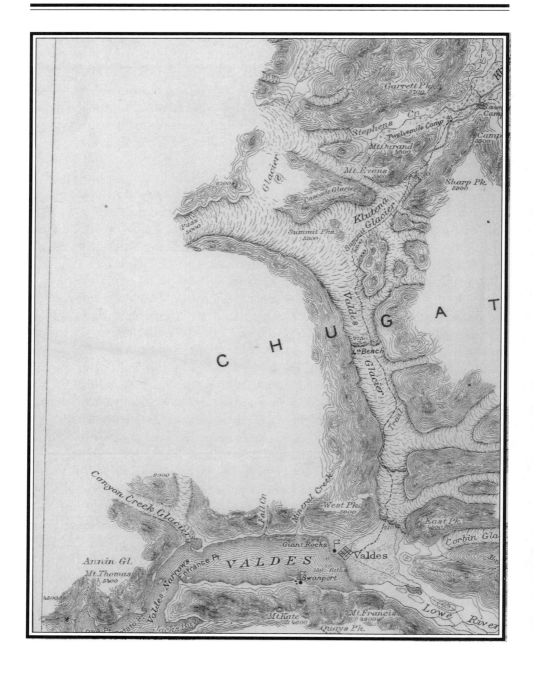

Valdez Glacier Trail from Valdes to Twelvemile Camp: from the Geological Reconaissance map of a part of The Copper River and Adjacent Territory, Alaskan Military Expedition, 1898, in Glenn and Abercrombie, Explorations in Alaska, 1899. Courtesy of the Valdez Museum and Historical Archives.

Chapter IV
Crossing Valdez Glacier

The Glacier Trail

For most of the two thousand people landing in April, Valdez was little more than a way station, a place to unload and temporarily cache their outfits before beginning the arduous haul over the glacier. Let us for now go along with the crowd and follow those 3,000 intrepid yet misguided souls who would defy death and danger to follow their dreams of instant wealth over the Valdez Glacier and into the unknown.

The glacier in those days, extended much farther down the valley than now to within five miles of shore. In the last hundred years it has retreated about two miles around an abrupt bend so that it is no longer visible from tidewater.

Before leaving "Camp Valdes," each person loaded his hand sled for the five mile trudge to the foot of the glacier. Joe Bourke describes the typical Yukon sled used along the Valdez trail.

> Our sleds: these are the most important articles we possess, for without them we could not move. We purchased them in Seattle for Six dollars each. They are of tough wood, strongly built and quite light for their size, being Seven feet long by Eighteen inches wide and Eight inches high. They are in skeleton form having Five slats for a top. They will bear a weight of at least Fifteen hundred pounds . . . . Our sleds served us well in the day time for carrying our loads and at night they did elegant service as bedsteads on which we placed our sleeping bags and we were thus kept up from the snow flooring of our tent. . . (Bourke, p. 8).

On the right hand side of each sled projected a 6 foot "gee pole" for steering; the pole projected upward and forward. The sleds could either be pushed from behind or pulled from ahead. Although Bourke remarks that sleds could bear 1500 pounds, they were never loaded that heavily. The steepness of the slope usually determined the weight carried. The average load was around 150 pounds. On a downhill run as much as 1000 pounds might be transported; whereas on a steep ascent less than 100 pounds was considered a load. Since the incline to the foot of the glacier was only 300 feet in a little over five miles, the sleds might be loaded with anywhere from 200 to 350 pounds.

For crossing the glacier and a summer's prospecting, each man required from 1500 to 2000 pounds of supplies consisting of heavy tarpaulin tents, a sheet metal Yukon stove for cooking and warmth (some carried oil stoves for the glacier), a

Well I have now learned what it is to make an ass of myself in earnest, one way at least, is to harness yourself up to a six foot sled, put 200 pounds on it and strike the trail for the foot of the glacier which is five miles away. Repeat this twice a day for a week and you soon have long ears . . . (Hazelet, 5/17/98). Photograph from the Bourke Scrapbook, courtesy the City of Valdez.

heavy sleeping bag, oil skins, extra boots, and lots of wool clothing, cooking utensils, axes and a whipsaw for boat building. Food consisted mainly of hard tack, beans, bacon, flour, rice, and occasionally powdered eggs, potatoes, and dried fruit. These he would load aboard his sled in appropriately sized bundles and begin trudging back and forth to the foot of the glacier. George Hazelet describes his experience thus:

> Well I have now learned what it is to make an ass of myself in earnest, one way at least, is to harness yourself up to a six foot sled, put 200 pounds on it and strike the trail for the foot of the glacier which is five miles away. Repeat this twice a day for a week and you soon have long ears . . . (Hazelet, 5/17/98).

Although the slope was not steep, a quick calculation reveals that moving one's gear was not an easy task — especially through newly fallen or slushy melting snow. Fifteen hundred pounds in two hundred pound loads would require at least seven trips back and forth to the terminus of the glacier — that is trudging through snow some seventy miles , half of them pulling a 200 pound sled. And this was only the beginning; once the foot of the glacier was reached, the performance would be repeated for another three hundred and sixty miles while gaining almost a mile in elevation. Even though a lightly loaded man might make the summit in two days, when the frequent storms on the glacier are factored in, it took many parties from six weeks to two months to lug their supplies across this formidable ice barrier. As R. F. McClellan put it, "this was something like work (Holeski, p. 308)." Guiteau observes that this prospect was too much to contemplate for some: "Every day we hear of some fellow we used to meet going and coming on the trail that has backed out and is offering his outfit for sale (Guiteau, *Diary*, 3/11/98, p.4)."

Gold seekers from warmer climes traveling back and forth on the trail were soon to learn that winter and cold were their friends and that those arriving earliest in the spring were to have the easiest time crossing the glacier. When the weather was clear, crisp, and cold, a good crust formed on the deep snow making sledding easy;

". . . the trail was all night lined with hardy prospectors and their various kinds of live stock, sledding and packing up the glacier with an energy and determination known to the gold seeker alone. (Schrader in Glenn and Abercrombie, p. 353)."

Photograph by Neal Benedict from the Messer Collection courtesy of the Cook Inlet Historical Society.

but during warmer periods, usually accompanied by heavy, wet spring snow or even rain, the trail became slushy and soft. Men and sleds would break through the crust or sink deeply into the soft snow. At these times, traveling at night when it was colder was the only solution. Later, when the spring rains began, many sections of the trail flooded; the glacier stream overflowed its banks, and small lakes formed at the terminus of the glacier. Later still, when the snow disappeared completely, it was almost impossible to transport the heavy supplies from the shore to the foot of the glacier without pack animals.

During winter and early spring, the glacier's many crevasses were filled with or at least covered by deep snow forming reliable snow bridges. But when the snow began to melt, many crevassed areas become impassable; and the snow-covered surface of the glacier was transformed into slippery glare-ice.

Although the constant trips to and fro fetching supplies could be monotonous and boring, there was still time for a man to reflect upon the natural beauty of his surroundings. Joe Bourke describes the trip to the glacier:

> Here nature presents itself in all its grandeur. We are traveling through a valley of an average width of probably Two miles. This is closed in on both sides by towering mountains covered with snow which frequently slides down into the valley causing a great noise in its descent (Bourke, Diary, p. 10).

Former school principal, George Hazelet was as much impressed by the human drama being played out on the glacier as by its natural beauty:

> The glacier is a wonderful thing. It begins almost at the edge of the sound at the time of year we reached it. And it extends in generally a northern direction for about 15 miles to its summit where it reaches a height of 5140 feet as indicated by our barometer, and crosses over the range it runs down on the north slope of the coast range for a distance of about 10 miles. . . . The man that starts in to pull 1500 lbs of supplies from Valdez to the top of the summit must needs

have plenty of grit, a good supply of muscle, for it is no easy task and I firmly believe that many of the parties now put down at Valdez alone to overcome this task would fail, but seeing others going over, with determination set in every line of their faces, they brace up and proceed and finally to their surprise almost, they have accomplished what under other circumstances would have proven impossible (Hazelet, Diary, 5/1/98).

Basil Austin describes the approach to the terminus of the glacier:

> As we obliquely approached the glacier, it surely looked like an impassable wall of ice, rugged and in most places almost perpendicular. Its face, hundreds of feet in height, exposed blue ice by the millions of tons, while snow covered and banked up on every place that was not too vertical for it to cling to. We could not see over it, but snow was drifting in whisps and swirls on the rim, although it was a clear day with no wind on the flat below (Austin, p. 23).

Glacier City

The camp at the foot of the glacier consisted of perhaps a hundred tents, some 300 people, and numerous transient caches. It was humorously referred to as "Glacier City." Joe Bourke remarks: "In speaking of this as a city it must be borne in mind that it is all under canvas and constantly changing but the change is not noticed for as one tent is moved out another takes its place (Bourke, Diary, p.13)."

If one were lucky enough to arrive in Glacier City just as another party had moved on up the glacier, it would not be necessary to dig down through the eight to ten feet of snow and tramp down a new tent site. Glacier City was the last opportunity to obtain fresh water from a nearby spring and a wood supply from a grove of cottonwoods two miles distant. As difficult as it was to add another several hundred pounds of wood to one's endless sled trips up and down the glacier, the fuel was essential to protect the men from the omnipresent cold hovering over the ice field — especially during the frequent severe storms. Hot meals were necessary for the morale of men performing the strenuous labor on the frigid glacier. More importantly, melted snow provided the only drinking water.

Surmounting the Terminus

Once the prospector sledded his supplies to the foot of the glacier, the first obstacle was the glacier's terminus itself or what were called "the first and second benches." Although Margeson describes the initial pitch as "a steep wall of ice," it appears to have been a steep scarp of broken blocks of ice stretching 150 feet at an angle of about 50° — not easily sleddable. Instead, the men devised a system consisting of rope, blocks and tackles, a technique which would serve them well on other steep benches higher on the glacier. Margeson describes the operation:

> . . . the first bench . . . is about sixty yards in length, and so steep that it was necessary to cut steps in the ice to get up. At the top we drilled a hole in the ice,

into which was set a post, attaching to it a pulley, through which we passed the end of a rope. Then attaching each end of the rope to a sled, about ten men would climb to the top, get hold of the rope and empty the sled there, and come down the incline, drawing the loaded sled up, carrying from six to eight hundred pounds of goods at a load (Margeson p.73).

Cooperation was the order of the day as men from different parties banded together to help each other over this difficult obstacle. Margeson remarks on the good spirit fostered by this cooperative endeavor:

> I consider this a remarkable thing, when men were working so hard as almost naturally to become irritable when very tired; and if anybody doubts that they were very tired, he would soon be convinced by a day or two of actual trial of it. Many hundreds were passing and repassing upon this narrow trail, and necessarily were often much in one another's way. And each one seemed ever ready to lend his neighbor any assistance needed, when it was possible for him to do so. (Margeson, p.74).

However, the good feelings were not to last. A steamer landed in the bay unloading 600 Swedes who immediately set about night and day carelessly hoisting their sleds over the bench beneath which the Connecticut Company was camped:

> During their first day's work at this place there was considerable excitement caused by a loaded sled breaking away when near the top, and coming down at lightning speed, ran over two or three men, quite seriously injured one of them, and crushing through one corner of my tent, where only a few moments before some of our men had lain asleep. We ran out and cautioned the men against such carelessness, and going in, sat down to breakfast; but before the meal was finished we were again startled by the cry, "Look out for a runaway sled!" and running outside, were just in time to see it upset and spill its contents in the snow just above us. This brought a sharp reproof from our boys, with another promise on their part to be more careful in the future (Margeson, p. 75).

Above the first bench, men with lightly loaded sleds (two men to a sled) followed a steeply switch-backed, half-mile course close in to the mountainside. The trail here exposed parties to the ever present danger of snowslides. This location was the site of a large avalanche on April 29th. Lieut. Brookfield recounts:

> When passing the second bench, we found that even the upper trail had been covered to a considerable extent in places by the snow slides of the 29th and 30th, and a number of men were then burrowing in the snow in an endeavor to find their buried caches. The first of these slides occurred on the evening of the 29th. It extended over to the trail on the side of the glacier, and had buried six men who had camped there at the beginning of the storm, together with some twenty-five burros which they employed in freighting. The men were all

taken out alive, suffering severe bruises. The burros were less fortunate, ten being killed and the remainder badly used up (Glenn and Abercrombie, p. 597).

Later in May, melting snows uncovered the bodies of three men who had perished in this avalanche.

The second bench was 225 feet rising at a steep angle. Blocks and tackles were again necessary. Above the second bench, men struggled with lightly loaded sleds up a steep slope another eighth of a mile before encountering the third bench. The second and third benches were so close together that they might be considered one.

A Final View of "Civilization"

The 1500 foot haul onto the third bench, the toughest yet, was usually accomplished in a series of three stages with block and tackle. Once atop the third bench, the men had surmounted the broken glacier terminus and were now on the glacier proper. From here, they encountered a steady, more gradual rise of 1,920 feet over the next eight miles. The top of the third bench appears to have been near the abrupt bend where the glacier suddenly trends almost due north as it does today. Margeson describes the scene from this vantage point 830 feet above the valley floor:

> While working upon the third bench of the glacier, the view down the valley was unobstructed. Valdez was in plain sight, and the trail leading to the glacier was marked by a long line of dark moving objects which were mere specks in the distance. They were men, pulling their goods up the trail on sleds as we had done. Every steamer or schooner which came into the bay could be plainly seen, and almost every day brought a new addition to the already large crowd of gold-seekers. We couldn't help thinking that if all of us were to strike luck, then indeed must Alaska be full of the shining metal (Margeson, p. 77).

As the top of the third bench was only a little over two miles from Glacier City, men spent their days relaying caches to a point 3 miles beyond the top of the bench (known as Five-mile cache or camp) returning each night to the foot of the glacier where wood and fresh water were plentiful. The tents and camping gear were the last items moved up onto the third bench. Some even pushed on farther up glacier to establish caches near the base of the fourth bench — a daily 20 mile round trip.

Five-mile Camp

Camps and caches quickly sprang up near the top of the third bench and all along the trail wending its way 8 miles up the glacier to the bottom of the fourth bench. The most populous was five-mile camp, so called because it was five miles from the glacier's terminus. Brookfield notes about 300 people camped here in late April. It was along this part of the trail that the first two deaths on the glacier struck in quick succession. The first and less dramatic occurred on March 22nd. T. H. Opdahl of Marshal, Minnesota died suddenly from heart failure while hauling his

"From the top of the Third bench we get a grand view of the surrounding country even to the Sound. We get a view of the valley to the shore and from this height the people wending their weary way over the trail looked like a prosession of ants. (Bourke, Journal, p. 24)."

Photograph No. B62.1.1343 "Third Bench Camp, Valdez Glacier, 1898" from the Crary Collection, courtesy of The Anchorage Museum of History & Art.

heavy load up the glacier. The second, and perhaps more tragic, occurred the very next day when young Leroy Murfin of Conger's Minnesota party was shot by the accidental discharge of a firearm. Murfin was lounging in his tent when another young gold seeker, D. Whirley of the same party, was rummaging through a pile of rifles in an adjacent tent. Unknown to Whirley, one of the 30-30's was loaded and discharged piercing both tents, fatally wounding Murfin who died the next day. Since Opdahl and Murfin were both Masons, members of that fraternal organization performed the funeral and burial. Fraternal organizations fulfilled an important function on the frontier assuming duties which ordinarily would have been those of friends and relatives back home. The bodies of both men were buried at the foot of the glacier not far from Glacier City.

Meeting the funeral procession on its way down the glacier, George Hazelet reflects on the sadness of death on this lonely trail so distant from family and home:

> It was a sad sight to see the poor fellow pulled down the glacier on the sled he had pulled so many days and so hard, now strapped to it, just as he had strapped his load. . . . About 50 men followed the body to the grave at foot of the glacier, no coffin, nothing but his blanket to protect him from the damp cold earth. . . . The poor fellow came into this wild country to find his fortune and found his death. Murphy [*sic*] was a young man and the news will be awful for his people to have. They [Murfin and Opdahl] lie buried close together, at the foot of a rugged mountain and there the glacier frowns down on them eternally (Hazelet, 3/29/98).

The sixteen-mile round trip to the foot of the fourth bench, necessitated virtually abandoning for extended periods the outfit on which each man's very survival depended. Under these circumstances, a rather unique honor system evolved. In short, thievery simply was not tolerated along the trail. New Yorker Joe Bourke remarks:

Both upon the summit and at its foot were temporary camps of consider-able size, and accompanying them large caches of supplies, of which the owners might well be proud, for they represented unusual value in the form of severe labor and hardship. . . . (Schrader in Glenn and Abercrombie, p. 351-52).

Photo fby Neal Benedict from the Messer Collection courtesy of The Anchorage Museum of History and Art.

To a person coming here from a city like New York the honesty displayed is perfectly astonishing. We go miles away from our goods and camp and nothing is ever disturbed. I have seen such things as pipes, gloves, knives and so on that were lost laid up carefully on the snow to one side of the trail so that the owner might get it on his return trip or in some cases the finder would take the lost article to his tent and there put out a notice requesting the owner to call for it (Bourke, Journal, p.12).

Whether the source of this remarkable honesty was a self centered empathy under survival conditions or fear of harsh repression is hard to say. It was probably a mixture of both. Writing about his arrival in Valdez in early March, Lute Guiteau, recalls forty-two years after the events:

It was with considerable shock that we observed our first case of Miner's justice. Our first contact with it was the sight of a man's body hanging from a tree. A little inquiry from a former California miner elicited the information that the death was the penalty in a mining camp for robbing another man's cache, and in that in Port Valdez, Miner's law was the only kind that seemed to be in effect. When we thought it over though, it seemed fair for in a camp it was necessary to leave all our goods in big piles, or caches, while we worked on the trail. If a man hadn't honor enough to leave another fellow's property alone, there wasn't room for him in the camp. The miners would usually give a man the shirt off their backs if a fellow needed it, but they certainly had no time for a thief (Guiteau, *"Golden Eggs,"* p. 11).

Guiteau, however, makes no mention of such an event in his diary recorded at the time. The March 8th entry in Horace Conger's diary seems to refer to the possibility of a second hanging: "About 900 camped here in the timber at the foot of the glacier. One man stole a sack of flour and I guess they hung him." The other diarists of the period fail to record a second hanging. After the publicity surrounding the lynching of Doc Tanner, it seems unlikely that the sensationalistic press of the period would have ignored a second hanging, but we find no mention of it. Despite the Hollywood image of the frontier, such incidents were extremely rare.

What seems most likely is that the Doc Tanner story became an oral tradition with a moral purpose. A slight twist to the story would provide a convenient warning to new arrivals. An early miners' meeting established the local penalty for thievery — anyone caught stealing $10 or less would have his goods confiscated and be banished; the theft of anything over this amount was considered grand larceny, and the penalty was hanging. Margeson certainly makes a connection between the hanging of Tanner and this ruling of the miners' "Law and Order Committee."

> This event [Tanner's hanging] caused the people to believe that crimes of greater or less magnitude would be committed from time to time, with such a conglomeration of citizenship during the year, that it should be impressed upon the minds of all who landed at this port that no crime, however small, should go unpunished (Margeson, p. 26).

One instance of alleged thievery did occur on the glacier. Bryan Pearson's *Journal* reads: "May 11 — Went to 4th bench to trial of man accused of stealing from a cache. Not proven but suspicious. Verdict of jury of 12 was that man have until Saturday to find man he claimed to have bought goods from and if he did not find the man he was to have till Monday to leave camp (Pearson, p. 185)."

The wisdom of this decision, which takes into account the ambiguity of the situation, lends credit to the miners' ability to administer justice even-handedly. Certainly, miners prided themselves in their justice system which was beyond the reach of civilized due process. Adam Powell recounts with his usual incisive irony:

> One Sunday, when Slate Creek was abandoned by all hands, because they were attending a miners' meeting in another gulch, I walked up the creek to find it deserted, and thousands of dollars in the yellow metal scattered around the tents in gold-pans and tin cups. No one was left to watch over the treasure, as thieves in such localities are not protected by law (Powell, p. 231).

Copper River Joe says of miners' courts:

> Cliques did not have their way here as there were too many total strangers to one another — so framing up, spite, greed and jealousy lost out. The only place you can gamble where these devils incarnate will lose is in a virgin and unsettled country, as these parts were then. . . Lawyers were not required or tolerated. . . . In the Sourdough's Moving Court, justice triumphs, and the Devil has no chance to secure a toe hold. . . . (Remington, p. 24).

He further observes, "Captain Abercrombie seemed inclined to protest the miners' rulings now and then, but of no avail. . . ." (Remington, p. 13). Although part of the military's mission was to establish order in the mining camps, Captain

Abercrombie was savvy of the American frontiersmen's resentment of government intrusion into civil affairs and except in one instance was loathe to interfere.

On to 12-Mile Camp

With a rise of about 100 feet to the mile, the sinuous trail up the long ramp from the third bench to the bottom of the fourth was not as difficult as the pitches which lay behind; but nonetheless was arduous work when hauling a sled loaded with 150 to 200 pounds. A round trip could easily consume an entire day, even though some sections could be sledded on the way down. Long days hauling the sleds back and forth even under cloudy conditions exposed the eyes to the sun's ultraviolet rays reflected from the snow. Despite the dark glasses they had purchased in Seattle, many began to experience a gritty, rasping sensation behind their eyelids as if a handful of fine sand had been thrown in their face. Numerous miners soon found themselves stumbling along the trail in broad daylight suffering from snow-blindness. Fortunately, sitting in a dark tent for several days provided a cure, but valuable time was lost for anxious men on their way to the gold fields. Miners soon discovered that carving a "bandit's mask" from cottonwood bark with narrow eye slits and blackened on the inside in Eskimo fashion proved more effective protection than their store-bought glasses. Likewise, the intense rays of the spring sun reflected off the surface of the snow resulted in "boiled lobster-like, skinned and peeled up red-onion like faces." (Remington, p. 14).

Above the Third Bench the trail made a huge curve to avoid an area broken into numerous crevasses. George Hazelet describes these hazards:

> The trail crosses wide cracks and crevasses which are perhaps hundreds of feet deep, scattered all over it are large boulders weighing thousands of tons waiting for the sun to melt the ice and snow and allow them to tumble down to the sea. A glacier is treacherous. It lies quietly waiting for its victims, with yawning cracks filled with loose snow. Once you fall into one of these cracks, without help near, and you have gone the way of all the earth. One might think they would remain forever in cold storage, but not so, for the treacherous thing is silently moving on to the sea. And while it might be years, yet in the course of time he would find that his bones would be mingling with the sands of the sea (Hazelet, 3/1798).

Although lives would be lost in these crevasses during the later, fall and winter mass exodus, it is a minor miracle that during the spring and early summer stampede of perhaps 3,500 gold seekers, completely inexperienced in glacier travel, not a single death would be attributed to this cause.

Because of the long haul to the fourth bench, the men were often caught in snowstorms and blizzards. At such times, visibility would be reduced to nearly zero,

"There were a few deep crevasses still open, but the trail naturally avoided them. Doubtless we passed over many that were bridged with snow, but we gave the matter little thought. The whole thing was awesome (Austin, p.27)."

"The snow that covered the crevasses had become too rotten to be safe, and those who crossed told of jumping cracks with spring-poles. If they had slipped they would have been put in cold storage forever, hundreds of feet below (Powell, p. 24)."

Benedict photo, Messer Collection, courtesy of the Cook Inlet Historical Society.

the wind chill would become horrendous, and all hauling would cease as men and women took refuge in their tents. Basil Austin writes:

> It was cold and windy. . . and we were glad to get the tent pitched and the stove going. . . .We conserved wood as much as possible... but must add that the sheet-iron stove was most economical; a small amount of dry wood working wonders in cooking and heating the tent. Just as soon as the fire went out the temperature dropped to that of the outside; one thickness of canvas providing no insulation (Austin, p. 25).

During extended storms, to conserve their precious wood supplies so laboriously hauled over the glacier, many would pass spend days huddled in their sleeping bags. When the storm abated, it was not uncommon to find tents, caches, and the trail itself buried beneath four feet of newly fallen snow. At these times, breaking trail proved a grueling task for the overly eager. But with over seven hundred people on this part of the glacier, the trail would soon be trampled down into a passable, hard surface. After these not infrequent spring storms, travelers would witness avalanches cascading down precipitous slopes. Bourke describes the view from the trail:

> Snowslides frequently occur from the surrounding mountains. The attention is attracted to the slide by the terrific noise which much resembles the sound made by a train of cars running into an inclosed depot. The view is beautiful. At first it resembles a waterfall with the fine snow pouring over the projecting rocks then it gathers the gravel and rock with it and down it comes pell mell to the valley. It would be dangerous were the trail near it, but we are at least a mile from the foot of the mountain and consequently we can enjoy the scene (Bourke, p. 25).

Still, heavy snow was much preferred to heavy rains. Hazelet records the following uncomfortable experience on March 30th.

It rained for first time since we came and tent leaked awfully. I awoke about midnight and found water had run into the bed to such an extent that I was compelled to pull the mattress up and pour the water off from the blankets. After a long time I went to sleep, again to be awakened about 3:00 AM with the feeling that my feet were in water. After due examination I found the water had run in and soaked through the tarpaulin and formed a pool around my feet. I got up, built a fire and sat up rest of night to get dry (Hazelet 3/30/98).

Animals on the Trail

While sledding up the long slope, there was ample time to observe one's trail companions — both of the two footed and four footed varieties. Almost all the recorders comment on the number and variety of pack animals. The most abundant were dogs in a variety of sizes and breeds, then horses, burros and mules and even a number of goats. Opinions differed as to which were the most valuable, although none seems to have really favored the goats.

Because their small hooves tended to break through the snow's crust, burros and mules, traditionally considered the prospector's friends, proved to be less useful than horses. Horses, while having some drawbacks, soon became the pack animals of choice — especially small western range horses. A horse could manage a 1200 pound load and could travel at a faster pace than a man pulling a sled. The preponderance of hand-sleds rather than horses on the trail was undoubtedly due in part to the expense of buying, transporting, feeding and maintaining horses in Alaska — although a horse could be purchased in Seattle for as little as $15. When it became apparent how valuable these animals could be on the glacier trail and especially in the interior, these same horses readily sold for as much as $450 in Valdez. The main reason, however, for the scarcity of horses on the Valdez Glacier Trail seems to have been misinformation received in Seattle. George Hazelet remarks:

> In the past two weeks a number of horses have been brought in, they are doing good work. . . . When we left Seattle we were told that horses could not live in this climate and it is possible that they will not live long. Feed is very scant and hay costs $100 per ton (Hazelet, 3/30/98).

Once their utility was established, horses were purchased on mid-glacier by larger companies such as Townsend's Keystone Company and Margeson's Connecticut Company. Prospectors like Hazelet and Meals, Benjamin Franklin Millard, Heber Smith and B. G. Leveroos would return to Seattle for horses for their next season's expeditions. Because the horses' hooves tended to break up the trail, rules established at miners' meetings restricted their use to the early morning hours when the crust was firmer.

Dogs are traditionally associated with hauling in the north. And, the newcomers brought with them dogs, lots of dogs — in fact so many dogs that Neal Benedict remarks: "Because of its colony of dogs, in the latter part of March, Camp Valdes was

"Some of the dogs seemed to feel the responsibility of their positions and to take as deep an interest in the progress of their loads as their masters did, pulling hour after hour as if their lives depended upon it (Neal Benedict, p. 171.)"

Photograph by Neal Benedict from the Messer Collection Courtesy of the Cook Inlet Historical Society.

a howling bedlam, especially at night; and their snapping and snarling, growling, barking, bow-wowing, and kiking in the forest or bare basswood timber near at hand was one of the features of the place (Benedict, p. 169)." Benedict further notes:

> The dogs were not confined to any particular species. There is a common impression among the uninformed on the subject that only Esquimaux dogs were employed on the northern trails, but this is quite erroneous; for St. Bernards, Great Danes, Newfoundlands, Collies, and a dozen other varieties, of all colors and sizes, were represented (Benedict, p. 168).

Apparently, some of these rather unlikely breeds performed quite admirably. Lt. Brookfield observes: "A large-sized, well-fed animal of the Newfoundland or St. Bernard species will pull as large a load as a man and will do fully as much work in a day, and they have the great advantage of being able to work in a soft trail, which would break through under any other animal (Glenn and Abercrombie, p. 596)."

The performance of some dogs was truly remarkable:

> Some of the dogs seemed to feel the responsibility of their positions and to take as deep an interest in the progress of their loads as their masters did, pulling hour after hour as if their lives depended upon it. When passing the fifth Bench, for instance, where the grade was about three times as heavy as upon the intermediate trail between benches, it was necessary to rest the dogs frequently. At a word from their drivers they would lie flat down in the soft snow at the side of the trail, with their mouths wide open and their tongues hanging out, as if almost totally exhausted. After a moment or two, at the word from the driver, they sprang back to their stations and pulled for about thirty feet more at the top of their strength when another breathing spell was in order (Benedict, pp. 171-72).

In their ignorance many brought breeds having little or no pulling instinct. In many cases, the dogs were mistreated through no fault of their own. Hazelet relates:

A man stopped at our tent today and got our gun to kill one. The condition we made was that he must take him out of hearing of the tent to kill him. Some men are so brutal. We see so much of it here, whipping their dogs when the poor things are doing all and more than they are able to, some will even ask you to give their dog a kick as you pass him. It is needless to say that we always decline with the remark that we are not in the dog kicking business at the present (Hazelet, 3/30/98).

The situation became so bad that a miners' meeting had to be called: "A few men were so cruel to their dogs, and beat them so much, that a committee was sent to wait upon them; and they were ordered to stop beating them, or they would be dealt with by the indignant miners. This seemed to have the desired effect, for from this time on the poor dogs fared better (Margeson, p. 79)."

Basil Austin describes an instance of rare poetic justice:

A noisy gang was camped near us. Nels called them the "Lion Tamers," as they had three dogs which didn't know just how to work with a sled. Con–sequently, they were always lashing and yelling at them. We strenuously objected on one occasion when they were being maltreated. Not more than an hour later, going down a steep pitch the man who had beaten them slipped and fell. The sled ran over him cutting or rather rubbing two good welts, one across his behind, the other above the knees. They more than matched in size those he had inflicted on the dogs. John said, too bad the sled wasn't longer and heavier. Anyway, he was minus a good pair of Mackinaw pants with what went underneath, and would be rather sore in the hind quarters for some time to come.

They shot the poor dogs while we were eating supper. While we were sorry for them, we felt they were better off than to be further subjected to such damnable treatment. No one had any trained native dogs. . . . (Austin, p. 41).

Cruelty was not always the case as Joe Bourke notes:

It seemed a commical [sic] sight at first to see a man or woman hitched to a sled with a dog making a team to draw a heavy load. For the dog there is no rest as on the return trip he pulls the sled while his companion in harness sits on and rides back. There was however one exception to this rule for one man was humane enough to put his dog on the sled while he pulled back on the return trip and the dog seemed to thoroughly enjoy it (Bourke, p. 11).

The People

Each reporter's reflections concerning his fellow trail-mates were as varied as those holding the opinions. They range from wonder and admiration to pity and contempt. Here is a sampling:

Joseph Bourke:

The best of order prevails everywhere. Every one is courteous to his or her neighbor and honesty is noticeable everywhere. Each seems to be intent on minding his own business. Here is to be met people of all nationalities and from all stations in life. They assemble here on a common level, the doctor, lawyer and laborer each taking his own share (Bourke, p. 11).

Luther Guiteau:

Its a great place to study faces and human nature. We passed almost every nationality on the face of the globe from a Siwash Indian to a Jap. . . . We met fellows today whom we noticed on the trail that looked jaded and careworn . . . But every one seemed good-natured and taking their medicine without a kick (Guiteau, 3/10/98).

They, two Virginians, were very fine young fellows, but too green to make a go of it on the trail. They knew absolutely nothing about taking care of themselves out of doors, or of camp cookery. They even carried with them several heavy canvas grain-sacks to be used, they told me, to hold the nuggets they had been told they could get for the picking on the Copper River bars! (Guiteau, *Golden Eggs*, p. 32)

Copper River Joe:

It will now be in order to mention that the human being is the most ornery and pestiferous contraption that Nature and others ever had to do with (Remington, p. 10).

Leroy Townsend:

People — Good class, generally, remarkably so, indeed. (Townsend, Letter March 20. 1898).

George Hazelet:

The trail is a great place to demonstrate character. Men trudge along with a big load behind them, looking as if they are pulling— their lives out, all the horrors depicted in their countenances, others go along quietly with a "Good morning" or a "how do you do" seemingly happy or pretending to be. Still others will whistle or sing some old familiar air, but there are few of this sort. The sled really seems to develop the worst side of a man, for most all are ready to scrap, on the least provocation (Hazelet, 3/30/1898).

There are people of most all nationalities but I think the Swedes predominate. There is the long lean man, the short fat one who puffs and blows like a porpoise as he tugs at his load up the glacier. There are old men, boys, middle aged men. Men with whiskers and men, a few, without them. . . . How I pity some of these people, tired, footsore and weary. They trudge along looking as if they could go no further. I think many might give up and drift back to where they came from (Hazelet, 3/30/1898).

Neal Benedict:

So far as this writer's observation went, there were no angels on the Valdes and Copper River trails, although there were some pretty fair specimens of humanity. The first four miles of toil brought most of the human nature in the toiler to the surface, and the loafers and bullies, cranks, weaklings, boasters and fault-finders had all thoroughly marked themselves for future identification before the first bench was reached (Benedict, p. 143).

William Treloar:

Here is a place to test men. A man's temper is tried. One soon knows the disposition of his partners. One may live neighbor to a person for years and never get acquainted with him, but work one week with him drawing a sled and packing until he's near exhaustion then go to camp together and eat there and tidy up camp and at the end of the week you will know him. . . . (Treloar, Memoirs, pp. 21-22).

There were many quarrels on the trail, a good many fistic duels and a good many parties divided but very little gun play which was remarkable for nearly everyone was armed but maybe the reason for that was that there was no booze. (Treloar, Memoirs, p. 22).

Addison Powell:

Possibly no more than one percent of them çould recognize a mine if they had camped on it for a year. (Powell, *History of Copper River Country,* Crary Scrapbook, Vol II, p. 205).

Frank Schrader:

Both upon the summit and at its foot were temporary camps of con–siderable size, and accompanying them large caches of supplies, of which the owners might well be proud, for they represented unusual value in the form of severe labor and hardship. . . . (Schrader in Glenn and Abercrombie, p. 351-52).

. . . the trail was all night lined with hardy prospectors and their various kinds of live stock, sledding and packing up the glacier with an energy and determination known to the gold seeker alone. (Schrader in Glenn and Abercrombie, p. 353).

Captain Abercrombie:

. . . . most of the men located in the various camps had probably never been out of sight of the smoke from a factory chimney that the hardships cncountered on entering the country were such as to turn back 75 per cent of practically every outfit . . . (Glenn and Abercrombie, p. 306).

My belief is there were not more than 10% of the 3,500 people who crossed the summit of Bates pass [Valdez Glacier] during March, April, May and June who did not firmly believe they could pan out a fortune and return to their homes in time to eat their Christmas dinner (Glenn and Abercrombie, p. 346).

Ninety-five per cent of them had failed in business ventures many times, and only joined the rush to the gold fields in the hope that they might be one of the lucky men to strike it rich (Abercrombie, 1900, p. 21).

Women on the Trail

The one oddity on the trail that fascinated most of the male diary writers was the presence of women. There appears to have been 20-30 married women and two single women from Boston making the crossing that spring. The men's attitude toward these women were strongly colored by the stereotypes of the time and varied from pity to genuine admiration for their "pluck." Many of the married women performed the traditional roles of cooking, cleaning and sewing while others pitched in pulling the sleds along-side their husbands — fulfilling the period honored role of wife as a husband's "helpmate" (Benedict, p. 150). One of these, a slim, well-dressed redhead, gained the admiration of practically every writer.

One woman, Mrs. Dowling, probably the wife of A.C. Dowling, became a living legend along the trail for her pluck, compassion, endurance and love of adventure. Copper River Joe describes her saving a party of prospectors retreating back over the glacier that fall. "When these men were about to give up, a woman by name of Mrs. Dowling, shamed them into a final effort and it saved them, because one among them had the nerve, though undoubtedly was the weakest physically (Remington, p. 22)." Conger notes that she was a crack shot. When Corporal Tully was struck with typhoid fever on the summit of the glacier, it was Mrs. Dowling who came to his rescue. Because of the reference to her marksmanship, it is probably Mrs. Dowling that Margeson describes thus:

There was one woman on the trail who felt at home at almost any kind of work, but seemed at her best when participating in some exciting adventure. She could guide a boat down the swiftest mountain stream equal to an Indian, and she seemed anxious for an opportunity to shoot the great Klutina River rapids. She could handle and shoot a rifle with the dexterity of an old hunter. She always joined her husband in his hunting excursions, and many birds and animals were brought down by her unerring aim. She was good natured and always jolly, having a pleasant word for every one she met, and seemed greatly to enjoy the kind of life she was living (Margeson, p. 82).

A peak above Klutina Lake is named in Mrs. Dowling's honor.

Margeson describes another woman who sounds suspiciously like Lillian Moore (or possibly Anna Barrett).

> One day, just on a tour of investigation, one of these women walked from Valdez to the foot of the glacier, then up the mountain of ice fifteen miles, and then returned to Valdez, making a round trip of forty miles in one day, over a trail which, at its best, was bad enough. This same woman walked over the glacier several times during the summer, and in her search for gold penetrated almost as far into the interior as did any of the men (Margeson, pp. 81-82).

Copper River Joe assures us of their morality:

> There were not many of the sporting class along the Valdez route to the Pot of Gold; neither gamblers or loose women, though there were a few adventurers of both sexes. The reason was evident for up to the close of 1899's prospecting season there were no gold strikes of importance. . . (Remington, p. 13).

Neal Benedict attributes the remarkable honesty and good will on the trail to the "refining" and "ennobling" influence of the women — a stock image of the period. He goes on to assure us that women were perfectly safe in this environment. "Every woman in the region was treated with respect and perfect consideration and the slightest wrong to her would have been resented by every man in the country as a special personal grievance. . . (Benedict, p. 151)."

Powell, on the other hand, highlights the shifting balance of power in family sexual roles accompanying the women's suffrage movement of the late '90s. He tells the following story:

> A small husband and his very large wife attempted to cross the Valdez glacier in 1898, by the man pulling a hand-sled and the woman guiding. When nearly exhausted, the little man sat down on the sled and, wiping the sweat from his face, said: "Mary, don't you wish you were back on the farm?" "No, I don't! It was Alaska, Alaska, if you could only get to Alaska you'd make your fortune; now, confound you, let's see you do it! Get in there and mush on!"
> And with a sigh the poor little fellow replaced the collar over his head and "mushed on." (Powell, p. 340).

12-Mile Camp

The encampment at the foot of the fourth bench, twelve miles from the terminus at an altitude of 2600 feet, was appropriately named 12-mile camp and numbered some 350 persons when Brookfield and Schrader visited in early May. The fourth bench rose at an angle of about 10° for about 1/2 mile then assumed a gentle rise for the next five miles before encountering the abrupt rise of the fifth and final bench or summit pitch. The fourth bench could be sledded with much labor without using block and tackle. On surmounting the fourth bench, Austin describes the disheartening view of the final summit pitch:

> From the top we obtained our first view of the Summit which made our hearts sink. About five miles distant and between two of the highest mountains was a white wall, appearing like a gigantic dam, reaching a long way up towards their peaks. Up the face of this was a dotted line representing the trail with men working on it (Austin, p. 29).

And indeed many became thoroughly discouraged by this point. Parties began to break up and men began to descend the glacier to Valdez to book passage home or to take part in the founding of the town. Margeson writes:

> At this time Messrs. S. J. Cone, T. O. Roggers, L. D. Hoy, and Dr. Kortright withdrew from the company, and returned to the States, making nine in all who had withdrawn since landing at Valdez. This greatly reduced our numbers, yet we pushed on as hopeful as ever that the future held a golden harvest in store for the plucky ones who deserved it (Margeson, p. 84).

Lost on the Glacier

One of the major reasons for discouragement was that the miners laboring so intensely and suffering so many dangers and hardships on the glacier were not sure that they were even on the right trail. There was little correspondence between the maps they had bought in Seattle and the route they were traveling (see the Allen map on p. 8). The US Coast and Geodetic Survey maps, based on Abercrombie's bogus report and Allen's authentic one, showed the glacier route running directly east from the end of Port Valdez roughly through Keystone Canyon to a lake and then down the Tasnuna river to the Copper. However, the prospector's route was definitely northerly. Hazelet early in his trip optimistically quotes Abercrombie's erroneous figures: "We will in the next two weeks have to reach the summit, and in order to do this we have to cross at least 12 miles of rough rugged glacier, now covered with snow. From the foot to the summit it rises at least 2000 feet (Hazelet, 3/17/98)." When Luther Guiteau descended the glacier to the upper Klutina River, he believed he would find himself on the Tasnuna or Tonsina Rivers.

It is thus ironic to find Abercrombie writing in his post mortem report on the Valdez gold rush to his superiors:

> After getting into the interior and finding the maps and guidebooks purchased from "fakirs" in the States to be wholly, unreliable, two-thirds of those who had successfully surmounted the glacier going in now became apprehensive lest their retreat over the glacier should be cut off by the melting snow (Glenn and Abercrombie, p. 346).

Near the summit, Margeson's party hears a rumor, probably perpetrated by those wishing to discourage competition in the gold fields, that the Klutina glacier route leads nowhere. They panic and send one of their party down the glacier to row clear out to Fox Island to interview Bill Beyer's native wife, Omelia, to find out if they were on the proper route. Their envoy soon returns with assurances from the native woman that the glacier is the correct route.

In late April, because of the uncertainty that they are following the proper route, the miners send a delegation to Valdez to the recently arrived Copper River Exploring Expedition. While Abercrombie remains discretely in the background, he dispatches Lt. Brookfield and Coast and Geodetic Survey Geologist, Frank Schrader, to survey the glacier route. They leave on April 26 but are turned back by a fierce five day storm. They finally reach the glacier's summit on May 3 recording the true distances and altitudes.

Location	Miles	Elevation
Valdes	0	0
Foot of Valdes Glacier	4	210
Top of third bench	8	830
12-mile Camp, foot fourth bench	16	2,750
Foot of summit	22	3,800
Summit	23	4,800
Foot of Klutina Glacier	29	2,020
(Glenn and Abercrombie p. 366)		

Miners' Meetings and Auctions

When parties broke-up acrimoniously, a miners' meeting would often be called to insure the equitable distribution of goods. As few wanted to lug their caches all the way back down the glacier to Valdez, a miners' auction usually followed. At these auctions goods could be purchased at a fraction of their cost in Seattle or Valdez. And no freight was charged for the long haul up the glacier. The cash economy was one of pure supply and demand.

An astute prospector struggling up the glacier realized that he could prosper by buying goods cheap and selling high. Copper River Joe describes "Oklahoma Bill" Hemple's decision to turn back:

> Bill must have thought there were better chances of making a living, and living longer if he got off the glacier so he said: 'Boys, I think I will go back — to Valdez and start a store, I know you will all help me when you strike it lucky, by getting your outfits from me and I will be coming in on one side all the time if I stay sledding.' Bill did make good in the store business, as all know that knew Valdez (Remington, pp. 15-16).

Hemple ran a successful general merchandise and out-fitting store and later founded the first bank in Valdez changing his nickname to "O.K. Bill." Bill prospered by taking advantage of the economy he observed on the glacier. Theodore Kettleson describes Bill's later operation:

> A well known character, Oklahoma Bill, operated a store with everything a prospector needed, including pack horses, for a much higher price than its value. Prospectors returning from discoveries, resold the equipment and horses to him after considerable amount of ranting language for much less than they paid him originally (Kettleson, p. 98).

Stormy Weather

As the rushers pushed their loads ever higher on the glacier, the storms became more frequent and more intense. On stormy days and nights, the omnipresent cold covering the glacier like a blanket became even more severe as wood supplies dwindled. While traversing this section of the glacier, Luther Guiteau was reduced to a final candle to heat a cup of coffee for a lonely guest who had wandered into his tent seeking refuge from the cold.

> When I told him how this prayer came to my mind that I used to hear when I was a kid by an old Methodist exhorter by the name of Van Rensler back in Freeport. 'When O God thou burnest up this world with fire; then O God we'll know thy power,' he said. 'The way I feel now O God can start his fire any time he wishes, for I prefer to be burned up, rather than frozen to death. . .' (Guiteau, 4/15/98).

Later, while camped beneath the summit pitch in one of these protracted storms, Guiteau remembers hearing a voice from out in the howling blizzard which he describes as belonging to either an "angel from heaven" or "a devil from hell."

> Last night just as we were ready to crawl into our sleeping bags, a big, burly-faced (half-Russian, half-Indian) man, wrapped in furs, stuck his head into our tent and asked in a roaring voice, 'Do you want a life-saver?'
> Philo said, 'What have you got? Come in.'
> 'What in hell do you suppose I've got?' the man replied. 'It ain't a load of coal — it's one bottle of Canadian Club whiskey, and it's the last one. Do you want it? Speak quick, for I want to get back out of this damned storm.'
> 'Yes, we'll take it,' we replied. 'How much?'
> 'Five dollars,' said the peddler, 'that's what I got for the hundred bottles I've sold along the trail already.'
> 'Where did you get one hundred bottles of Canadian Club in this God-forsaken country?' we asked.
> 'Oh, I got it from the purser of a big boat that just landed at Valdez, and when I bought it, he said for me to distribute it along the trail up on the Glacier near the summit, for he had heard there was a lot of fellows up here who'd been snowbound fer a week and he 'lowed as this 'ud do you boys some good.'
> 'Say, Mr. Lifesaver,' one of our boys asked, 'how did you get up here,

and when did you leave Valdez? Valdez is twenty miles from the summit, and we're a thousand feet below the top on the inside!'

'How did I get up here? That's a hell of a question to ask! Stick your heads out of the tent and you'll see how I got up here. Out here are four big Alaska huskies hitched to my sled, and I left Valdez at five this evening, and it's now a little after eight. Them dogs will take me any place where there is snow and ice. Say boys, I've been offered a thousand dollars for that leader, and believe me, he knows the trail. When I crack the whip, you ought to see them stretch the tugs. Well, good night boys, and I'll be back to Valdez in an hour and a half!'

Now, it seems peculiar that such an incident should make such a profound impression on me, but from the space allotted it both in my diary and in my memory, that bottle of whiskey, coming like a gift from heaven after weeks of struggle, was one of the outstanding bright spots of my experiences on the Valdez Trail (Guiteau, *Golden Eggs*, p. 33).

Summit Camp and the Great Storm

Approaching the foot of the summit in late April, one would note a kind of traffic jam created by delays in overcoming this last steep and difficult pitch rising over a thousand feet in a mile. Rows of tents and caches lined the trail on both sides for a good 3/4 of a mile right up to the base of the summit. Goods had to be relayed in stages all the way up this steep incline by block and tackle occupying several weeks for some parties — especially when work was delayed by the frequent storms.

On April 26, it began to snow in a way that only someone who has passed a winter in Valdez can even begin to imagine. For a full five days the storm raged around the little settlement dumping prodigious amounts of snow. The men dared only step outside briefly to shovel the heavy snow off their tents so that they would not be buried alive. The rest of the time, they snuggled in their sleeping bags and listened to the howling winds outside; for most of their wood supplies were gone. Windlasses, packing crates, and tent stakes were burned when available. Firewood, if it could be purchased at all, sold for 25¢ a pound and kerosene for $5 a gallon. By the time the storm began to abate on April 30, men emerged shivering from their tents to discover 7-8 feet of new snow on the ground and often 12 feet where drifting had occurred. Not a cache nor a tent was visible in any direction.

Avalanche

Of our diarists, Hazelet, Conger, Guiteau, Townsend and Austin were all fortunate; for by April 30th, they had already descended Klutina glacier down into the promised land. Likewise, the military had prudently retreated off the glacier to Valdez at the beginning of the storm. However, Margeson, Benedict, and Bourke were all camped at the bottom of the summit on the night of April 30th.

"It had begun to snow and a storm set in in earnest. The snow flakes [were] so large and came down so thick we could hardly see our neighboring tents only a few feet from ours. . . Well it snowed and snowed and we shoveled and shoveled till we were throwing snow ten feet high so we finally raised our tent. Nights we would build up piles of sacks and boxes under the ridge pole to keep the snow from breaking out tent. In the morning we would have to shovel our way out of the tent then clean the snow off."(Treloar, Memoirs, storm end of April, 1898, pp. 26-27).

Photo by Benedict from the Messer Collection courtesy of The Anchorage Museum of History and Art.

At 10 pm, it began as a slightly muffled swoosh high on the mountain peaks to the east of the camp, then gradually accelerated to a deafening roar as millions of tons of new snow cascaded down the steep mountainside tearing away huge boulders in its advance. When this snow mass reached the glacier floor it transformed itself into a gigantic snowplow that fingered its way out onto the middle of the glacier sweeping away caches and smothering tents in its line of advance. About twenty tents were buried beneath the snow which was six to fourteen feet deep. Some tents were buried while others not six feet away stood unharmed.

The reaction of those untouched by the avalanche was swift and heroic. Half-clad and barefooted, they grabbed their shovels and rushed out into the cold night air working frantically to locate and free those buried in their tents. The swift action succeeded in rescuing 25 of those buried in the slide. However, two men, B. Van Antwerp and Joseph Fournier, were not reached in time and died of suffocation. A third man, Joseph Thiery, in the same tent escaped alive. Joseph Bourke rescued a man named Johnson who later died on the steamer trip home from injuries sustained in the slide. After the avalanche, the men moved the tents to a safer location farther out on the glacier — which proved fortunate for a second avalanche cascaded down the slopes early the next morning burying a number of caches but no tents. Still, a third snowslide occurred the next day but without any losses. (Bourke 34-35).

The miners convened a meeting and selected a coroner, C.B. Smith, secretary. Joseph Bourke and jury of twelve. Following the proper legal procedures, testimony was duly taken from several witnesses, and the verdict of death by suffocation was delivered. Since the two were both Odd Fellows, members of that organization performed the funeral which was conducted near the foot of Klutina Glacier on the interior side of the mountains where the two were buried.

For weeks after the slide, miners with long poles could be seen patiently probing the deep snow looking for their lost caches. Since many had been shoved

Shorty and Jack: "[Jack] was buried under a snow-slide and his interests were neglected for eight days, which he passed in sad reflection on the fickleness of human love, doubtless imagining, in his ignorance of the true cause of his imprisonment, that he was being studiously punished by his master for some entirely unknown lapse of duty on his part. At last it occurred to "Shorty" to dig the dog out and he immediately did so with the result of rescuing Jack still alive, whereat his dogship was of course becomingly thankful; but some how he was not over demonstrative in protesting the old undying affection for his very deliberate master (Benedict, p. 171)." Photo by Neal Benedict from Messer Collection, courtesy of the Cook Inlet Historical Society.

over a great distance by the advancing snowslide, they were never to be recovered. For many, the loss of their outfit was the last straw, and they turned around and headed for Valdez. Others, undaunted, purchased whole new outfits from those turning back and continued over the glacier.

A remarkable story made the rounds of the camps. Eight days after the avalanche, some miners were probing the snow for lost caches when one heard the faint whines of a dog beneath the snow. They dug down and lo and behold Shorty Fisher's dog, Jack, emerged from a burrow in the snow where he had been imprisoned for the last eight days. Although ravenous, Jack seemed little the worse for wear and in several days was back on the trail hauling his load up the glacier.

The Conquest of the Summit

The final challenge of surmounting the steep summit pitch with the tons of necessary supplies was the most difficult of all. Margeson describes the struggle of the Connecticut Company to raise its 15 tons of provisions to the top.

> At last we had all our goods at the foot of the summit, and the next move was to get them upon the top. We established relays of men up the summit, and one man would take a sack of flour and carry it a hundred yards, when another man would take it, carrying it another hundred; then a third, and fourth, and so on. In this way one hundred sacks were moved to the top, when this plan was abandoned, and the rope and pulley again brought into use. We stretched one thousand feet of rope, and sent fifteen men to the top, who seizing hold of one end of the rope, came down, thus drawing up a sled load of ten to twelve hundred pounds on the other end.
>
> We would haul all our goods up thus far, plant our stake another thou—sand feet up and repeat the operation. It took five pulls of this sort to get our goods from the foot of the summit, or the last bench, to the top of the glacier (Margeson, pp. 101-102).

Summit Camp: "[We] pulled on to the foot of the summit 5 miles away. Good weather and a good trail favored us, and we reached there about 12 p.m. . . . This is the hardest proposition of the pass as we soon found. The ascent is at an angle of about 45° and it is about 3/4 mile long. From the trail a short distance back the packers remind you of nothing so much as flies on a smooth white wall. (Townsend, Letter, 4/22/ 98)."
Photo by Neal Benedict from Messer Collection, courtesy of the Cook Inlet Historical Society.

At last the bone-weary climbers stood atop Valdez Glacier's summit; not ten miles below, the timbered headwaters of the Klutina Valley hove into view. Most suddenly experienced a sense of accomplishment and exhilaration breaking into biblical comparisons. George Hazelet marvels "Moses-like we, Jack and I, have been to the top of the Mountain and viewed the promised land (Hazelet, 4/9/98)." Floridian, Neal Benedict speaks of the timbered land below as "God's Country" while Lute Guiteau refers to it as "A garden of Eden." The young, English immigrant, Basil Austin, remarks, "I believe it was Joshua who had sent scouts ahead to view the Promised Land. I remembered a picture in our old family Bible of the two scouts with a pole over their shoulders and a huge bunch of grapes suspended between them. As a kid, I figured they had the 'milk and honey' in their pockets. We heard similar reports of Timber Camp down on the other side. The grapes were represented by spruce trees, and the 'milk and honey' perhaps by real water from a creek and plenty of dry firewood (Austin, p. 32-33)."

Standing on the summit, George Hazelet took one last backward glance down the glacier towards Valdez: "It was a wonderful sight to stand on the summit of that glacier 5,000 feet above the sea and look back 18 miles to the coast and see the black streak of humanity winding its way over the snow and ice like a huge snake might crawl. Going where? God only knew. Many to their death and all but a few to sad disappointment (Hazelet in Towers, *Alaska History*, p. 33)."

Finally the Cavalry Arrives — Late

The Copper River Exploring Expedition under Captain Abercrombie arrived in Valdez aboard the *Valencia* on the 18th of April with instructions to survey an All-American route from Port Valdez to the Yukon. Captain Abercrombie immediately dispatched several parties to survey possible routes to the interior. Lt. Percival Lowe was sent up the river that now bears his name to attempt to find the rumored Russian

"I was subjected to the miserable humiliation of being passed by the civilians with their outfits, having in use horses, mules, burros, dogs, and even goats. Our Government party was, of all the outfits I saw, the most poorly provided with means and material for progress. I observed that range horses, half-breeds of 800 or 900 pounds in weight, such as are raised in Montana and Wyoming, are the animals of the greatest utility upon the trail, and best retain their flesh (Preston in Glenn & Abercrombie)."

Photo by J.E. Thwaites, No. 1280, from the Greely Collection, courtesy of the Alaska State Library.

trail. Lt. Brookfield was dispatched over Corbin Pass and Lts. Preston and later Brookfield and Schrader and then Rafferty were sent to explore the glacier route being pursued by several thousand prospectors. By late April, Lt. Brookfield acknowledges "that the leading prospectors had found the right trail to the Copper River (Glenn and Abercrombie, p. 597)." However, Abercrombie back at camp strangely concludes, "As the men were beginning to fag out, I considered that I had taken every justifiable means to penetrate the Coast range by man power (Glenn and Abercrombie, p. 565)." Apparently, in the Captain's estimation the U.S. military without pack animals was not capable of coping with conditions on the glacier now being faced by several thousand civilians including women.

Abercrombie leaves in late May for Seattle to secure pack animals and returns in early July with twenty semi-wild mustangs. He gives Lt. Lowe the pick of the best eleven pack animals and instructs him to cross the glacier with three men and attempt to reach the Forty-mile River via Mentasta Pass. Lowe, after two failed attempts on July 8th and July 10th, finally succeeds in crossing the glacier on July 13.

By now, Abercrombie is getting desperate; once again he might fail to accomplish his mission. It is now so late in the season snowbridges covering crevasses had melted making the glacier impassable for pack horses while the Lowe River was swollen by heavy summer precipitation and glacial melt water. He writes:

> On consulting with Lieutenant Lowe it was decided that the course of duty lay over the glacier, even if all the stock and some of the men were lost in making the attempt. As it was a last resort, and utter failure stared the expedition in the face if the passage of the glacier was not successfully effected, I decided, after the return of the men sent over the Bates Pass [Valdez glacier] with the Lowe expedition, to organize and equip a second expedition of four sections.

> Section No. 1 was commanded by myself, its object being to cover as

much as possible of that region embraced in what was known as the head of the Copper River district.

Section No. 2 was placed in charge of F. C. Schrader, geologist of the expedition, to explore and map the district embracing the Archer (Tonsina) and Teikell rivers, flowing into the Copper River from the west and heading in the Coast range of mountains back of Port Valdez.

Section No. 3 was placed in charge of Emil Mahlo, topographical assistant of the expedition, whose district embraced the Konsena and Tasnuna river valleys.

Section No. 4 was placed under the command of Corporal Robert Heiden, Fourteenth Infantry, whose orders were to cut and grade the trail up Lowe River and over Keystone Pass, thence to Thomson Pass (so named in compliment Hon. Frank Thomson, of Pennsylvania), connecting with sections 2 and 3 on the head waters of the Tasnuna and Archer rivers. (Glenn and Abercrombie p. 566).

Lowe then sends word down the glacier that without a snow pack, he does not consider the glacier feasible for the horses. Abercrombie once again vacillates. He decides to force his way up the Lowe River. However, after being swept off his horse in trying to ford the raging river, he decides that perhaps the glacier is the better route after all. Now it is the end of July, and he has accomplished nothing. Fog and rain have settled over the glacier further delaying a start. Finally, on August 5, despite the constant rain, Abercrombie orders the expedition to proceed up the glacier. Addison Powell and Lillian Moore were among those who accompanied him.

To surmount the glacier terminus, the party led the horses up through a maze of crevasses. The constant rain made the ice so slippery in these fissures that it was not uncommon for a horse to slide some forty feet before coming to a halt. Steps had to be chopped into the ice for the passage of man and beast. The steady rain took its toll not only on the party's morale but also its food supply — "the humidity was so pronounced and so continuous that bacon and ham became one mass of mold while the water of crystallization in the sugar being liberated the sugar wasted away in the form of a syrup (Glenn and Abercrombie, p. 304)."

Once on the terminus, it was necessary to lead the horses carefully past steep icy slopes and yawning crevasses. Missteps and mishaps were inevitable. Lillian Moore writes of the horses:

> It was frightful to see them just hanging by the fore feet over a crevice 2000 or 3000 feet deep. One slip and would be goodbye and they seemed to realize it. One poor horse's foot slipped on a ridge, he sat down, could not get his footing, so just turned his head, looked down, and deliberately rolled himself over and over back, and down 200 feet or so; he turned over at least a hundred times. When he got to the bottom he laid there a minute, shook himself, got on his feet and started up again (Lillian Moore, Letter. p. 1).

Fog then settled in over the party limiting visibility to some ten feet. The guides soon became confused and were constantly getting lost. Still the military continued chopping steps up the steeper, slippery ice pitches for both horses and civilians. At this point Captain Abercrombie with his usual contempt for those not in the military exercised some of his rather unique leadership qualities: ". . . I packed a 5-gallon keg of whisky, to use as a stimulant during the night and the following day. . . some of the civilian employees were inclined to be a little weak-kneed, as it looked very much like climbing up the fog into space. But by a judicious use of the stimulant referred to, this was overcome (Glenn and Abercrombie, p. 307)."

By 9:00 p.m. the expedition had reached 12-mile Camp at the bottom of the fourth bench. The incessant fog and rain had changed to a mixture of rain, sleet, hail and snow and the glacier wind began to blow a gale. The men were forced to pitch their tents on the bare ice. Moore notes the tents were of little solace as no blankets were provided so most spent the night tramping back and forth to keep warm. Although the captain had failed to provide his men with blankets, he did provide them with a liberal supply of the above mentioned stimulant both that evening and the next morning. After a miserable night, the expedition surmounted the fourth bench where it encountered a dangerous crevasse field covered with rotten snow bridges:

> These snow bridges which arched the deep crevasses were in a decaying state and liable to slump down at any moment, especially when disturbed by the weight of man or horse in crossing. The span varied from a few to nearly a score of feet, with the crevasses descending hundreds, in some places pro–bably thousands, of feet in depth (Schrader in Glenn and Abercrombie, p. 355).

> Now and then the horses would plunge and drop through one of these treacherous bridges, but before proceeding far would catch themselves with their fore feet on the ice. Then the pack would be undone, the ropes placed around the animals, and by this means they were saved (Koehler in Glenn and Abercrombie, p. 607).

Amazingly no horses were lost, however, many were severely bruised and lacerated sometimes leaving a trail of blood behind them. Unfortunately, all but one of the horses starved to death that winter. Several years later, Lillian Moore estab–lished a rescue mission for prospectors' horses.

The usual summit storm was blowing when they passed this point before descending down into the Klutina valley. Unlike earlier prospectors, however, they found a restaurant tent here serving tea and coffee. Abercrombie's late summer crossing of the glacier with loaded pack animals in only 29 hours was really a remarkable achievement defying the conventional wisdom that it could not be done. However, it is hard to attribute the significance he later ascribes to it. "The arrival of the expedition with the pack train at Copper Center effectively checked the stampede by giving a practical demonstration that where a pack train could go man could also travel with safety (Glenn and Abercrombie, p. 346)."

Chapter 5
Coming into the Country —
Down the Klutina

Descending into the Promised Land

Compared to the trials experienced on the oceanward side of the glacier, the final descent from the summit down the eight mile tongue to the Klutina River valley was a relatively easy task but still not without its hazards. Most commentators were pleasantly surprised at the stark differences they encountered once passing over the Chugach range. The steep mountain barrier causes the stormy, damp air from the Pacific to rise, cool and wring out its moisture onto its southern and western flanks. It also acts as a great wall holding back the warm, moist air on the ocean side from the drier, cooler air of the interior.

Because of the abundant snowfall, the glacier on the oceanside stretches some 18 miles from the 4800 ft summit down nearly to sea level. However, on the drier, interior side, it is only eight miles to the terminus at the 2,300 ft level. Timberline began four miles beyond. The descent was from winter into spring, from a swirling world of wind, clouds, snow and rain to one of almost perpetual sunshine. In addition, the gold seekers were finally hauling their loads downhill and could sled from 600 to 1000 lbs. in a single trip.

When the snow was deep, some pulling was still required despite the downward slope which, although not as severe as on the southern side, still had several relatively steep benches. However, for prospectors traveling on a slick, well-packed trail with a heavy load, the hauling could quickly become downright exciting. Men would dig in with their "gee" poles to act as brakes; this failing the only recourse was to veer off the trail into the soft snow usually resulting in an upset and spilled gear. Copper River Joe, while telling a bit of a "stretcher," captures in his colloquial style if not the facts at least the spirit of these adventures:

> The incidents arising from this down-hill slipping were numerous, humorous, maddening, and pathetic, as some of the mushers could be seen down in the canyon-like crevasses, wringing their hands over piles of kindling wood that were once boxes; and over piles of scattered dried fruit and beans and broken packages. The most humorous stunt that happened during Joe's descent down this infernal slide was that of Doc Dinsmore, a physician who left a good practice in Peoria, Ill. Doc loaded on about all he could, as he wasn't going to use a drag pole and get run over like Joe did. So Doc mounted his sled belly bumper and turned her loose. The snow flew and gravel also, from the trail, with boulders now and then. Also the mushers coming up for other loads hopped about lively in keeping away from the juggernaut. One of Doc's soles

on his moccasin shoe pacs commenced smoking and flopping loose, from trying to steer with his foot, and about the time the other shoe pac sole began to flop, Doc lost his cheery confidence, and when his big toe- slipped he lost his grip, then with the yells that came from the mushers to "stay with her, Doc", and "fall off", "fall off, you damned salmon head, or you'll get killed", he took a tumble, falling off alright, when leaving the trail at the first sharp turn, but his sled kept on, making a flying leap off the cliff of the moraine and dropping thirty feet down onto some big boulders, adding another heap to the Argonauts' kindling wood pile, to scald his first pot of coffee on terra-firma for over a month. "Bravo, Doc," they yelled, "but there's many a slip twixt the Devil's Slide and the Glacier's Hip." (Remington, pp. 19-20).

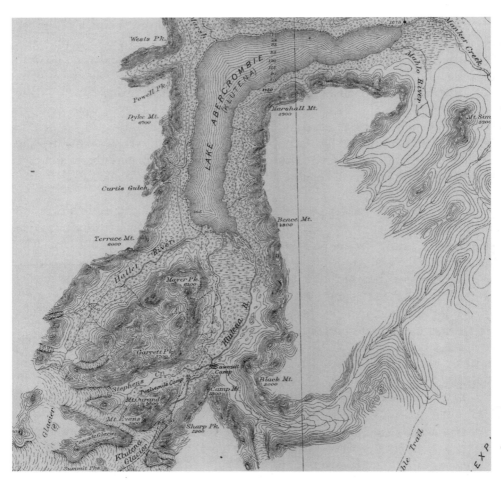

The Summit to Lake Klutina. From the Geological Reconaissance map of a part of The Copper River and Adjacent Territory, Alaskan Military Expedition, 1898, in Glenn and Abercrombie, Explorations in Alaska, 1899. Courtesy of the Valdez Museum and Historical Archives.

Boulder Camp rested on the moraine at the foot of Klutina Glacier. Most crossing the glacier moved on to Twelve– mile Camp where there was a ready supply of wood and water. However, crossing back to Valdez, this was the closest camp to wait out inclement weather on the glacier.

Photo by Neal Benedict, in the Messer Collection, Courtesy of the Cook Inlet Historical Society.

Once off the glacier a few set up tents immediately at "One-mile" or "Boulder" camp, but most headed along the headwaters of the glacier stream or upper Klutina River to the first spruce groves about four miles downstream. Here, they established a camp on the left bank called "Timber" or "Twelvemile" camp since it was supposedly 12 miles from the glacier's summit. In mid-July when Lt. Lowe passed, three hundred prospectors were camped at his point. Men suffering from six weeks of sleeping on glacier ice, being constantly damp from perspiration or the persistent precipitation, chilled to the bone, having to melt snow for drinking water and subsist on a steady diet of beans and bacon luxuriated in their new surroundings. There were spruce boughs for beds; firewood for dry warmth; water could be conveniently dipped from a nearby stream; and fresh trout abounded in the creeks; while plump rabbits crisscrossed the trails and ptarmigan filled the air. Later on, there were abundant greens for the gathering, and edible berries weighed down the bushes.

But alas, few earthly paradises are perfect. For it was here that the prospectors were to make their first acquaintance with the Alaskan mosquito which was to be the major complaint throughout the rest of the summer. Addison Powell describes his experience: "Our ascent of the Copper River was the same old story of a battle with gnats, flies and mosquitoes. It is very probable that many of those mosquitoes could whip a wolf. They are the embodiment of bravery. I have seen, a single mosquito attack a full-grown dog (Powell, p.226)."

In late April, while Hazelet and Meals were camped at Twelve-mile, a party returning from the interior to Valdez stopped at their tent. The leader of this party was none other than Powell's 72 year old confidante, Captain I. N. West. Because of illness and the marshy terrain west of Klutina lake, West was unable to push through his trail northward to the Chistochina as planned. West tells his hosts that he had

"At the place where the old man was drowned a sawmill has been established. The machinery was brought in over the glacier and set up on the bank of the stream about Five miles from the glacier. It Is almost a primitive affair. It runs by steam power and was constructed and Invented by the present owner. It did its work well. The charges were Twelve cents a foot for lumber but anyone furnishing their own logs got them sawed for Eight cents per foot. It was not a paying concern as most of the people did their own whipsawlng. (Bourke, p. 51)." Photo by Neal Benedict, Messer Collection, courtesy of the Cook Inlet Historical Society.

previously been to the headwaters of the Copper River and there was lots of gold there. Most probably, Hazelet and Meals like Powell believed West's story for they headed directly for this area and eventually discovered West's gold on the Chistochina and Slate Creek.

The Upper Klutina

Basil Austin, arriving early in the spring at Twelve-mile camp, had an easy time of it. The glacier stream was little more than a large trickle and the still abundant snow provided easy sledding the thirteen miles to Lake Klutina. However, by mid-May the melting snow-pack had transformed the little glacier stream into a roaring torrent several feet deep and 100 feet wide. Later arrivals had to stop and construct boats from the abundant timber, floating their goods down to the lake.

Boats built at Twelve-mile were of every possible type and quality of construction from crude rafts to elegant looking craft which were "fine enough to grace the waters of any aristocratic summer resort (Margeson, p. 117)." Some were long, graceful bateaux while others were stubby, flat bottomed scows varying in lengths from twelve to thirty-six feet. Most were in the twenty to thirty foot range. Slow, careful craftsmen like Jack Meals and Basil Austin built sturdy boats which proved worthy of the pounding they were to receive; while others, who had little skill in carpentry, hammered together craft which looked more like oversized packing crates than boats. These soon disintegrated in the rough and tumble rapids.

Some dragged "knock-down" boats of steel, canvas and wood across the glacier to reconstruct in the interior. Most brought hammers, nails, oakum, and whipsaws and constructed boats from spruce trees employing the pitch for caulking. Margeson estimates that over a thousand boats were built on the Klutina that summer. If hauling a sled over the glacier was grueling work, whip-sawing a 20 foot long log into planking proved equally so — plus being a phenomenally boring way

"From Timber we moved down– stream about half a mile to where the water was deeper and the timber better and here we built the Koo Koo. . . The building of boats put an end to sledding for this season and to tell the truth we were not sorry for it. (Bourke, p. 38)."

Photograph by Neal Benedict, Messer Collection, courtesy of the Cook Inlet Historical Society.

to spend a day. An enterprising outfit from San Jose sledded a portable sawmill over the glacier and set it up three miles below Twelve-mile at "Sawmill Camp."

> It is a most a primitive affair. It runs by steam power and was constructed and Invented by the present owner. It did its work well. The charges were Twelve cents a foot for lumber but anyone furnishing their own logs got them sawed for Eight cents per foot. It was not a paying concern as most of the people did their own whipsawing (Bourke, p. 50).

Although the glacier stream of the upper Klutina was little more than a swollen creek, it proved to be a treacherous body of water. Its depth rarely exceeded three feet, but the strong current and heavily silt-laden water could easily sweep a man off his feet. By midsummer, the creek accounted for at least two drownings: L. J. Mytinger drowned in the manner described above, while Charles Kelly died when his jerry-built boat overturned in the swift current. A third man, August Grund, a member of Conger's party, died of scurvy along this same part of the trail on May 26th.

The most frequent accidents on this stretch of the river, however, were parties wrecking or overturning their boats. Diarists report numerous prospectors drying out salvaged, life-sustaining supplies all along the stream banks. The swollen stream overflowing its banks contained many hazards such as submerged boulders, snags, sand and gravel bars, tree trunks and indeed whole trees growing in midstream. A boat thrown broadside against these obstacles was easily capsized. As the stream pursued its crooked course, the swift current darted back and forth from bank to undercut bank where tree roots and limbs reached out to tear at the rapidly careening craft. Even if one were to avoid disaster during the frequent groundings, it was often necessary to climb out into the icy water to shove the boat off thus leading Copper River Joe to dub the stream "Wet Seat Creek."

It was with more than a sigh of relief when the prospector finally landed at Peninsula Camp at the head of Lake Klutina. As its name implies, this camp occupied

"It may be judged what a nuisance it is to be run up on one of these gravel bars when I state that the water is icecold. The gravel on which we have to walk is of cobble-stone size and we were all barefooted, having taken off our shoes before starting so as to be prepared for a compulsory swim in case of accident. (Bourke, p. 60)."

Photo by Neal Benedict, Messer Collection, courtesy of the Cook Inlet Historical Society.

a 500 yard wide 40 foot high peninsula jutting out into lake Klutina. Because the peninsula was exposed to the glacier winds and bare, having been previously burned over, it was relatively free from mosquitoes. During the course of the summer, an average of about 50 tents here accommodated traffic in both directions. As the summer progressed, more and more miners became discouraged, auctioned off their supplies, and began making their way back over the glacier to Valdez. By midsummer, there were as many leaving as arriving.

Klutina Lake

At this point, most of the earlier arrivals including the military believed that, as their maps indicated, they were on the Tasnuna River and that they had arrived at Abercrombie's mythical Lake Margaret. All who witnessed the lake seem to agree that it was an earthly paradise. Joe Bourke describes the setting: "The Lake is a most beautiful sheet of water Twenty four miles long with an average width of Four miles. The banks are fairly regular with a dense growth of timber down to the water edge. The scenery is beautiful and the receding mountains on either side lend a charm to the view (Bourke, p. 45)." To which Lillian Moore adds: "When I got to the head of the Lake it was as far as I wanted to go. It is a beautiful place and is called 'God's Country'. Over there it never rains. The mountains are covered with flowers and berries. You never saw such large currants, black and red raspberries, blueberries, strawberries, all kinds of fruit; the loveliest sweet peas and forget-me-nots: and talk about scenery — it is something grand! (Moore, Letter)."

Both Margeson and Bourke remark on the pleasures of boating on Klutina Lake. All manner of craft plyed these waters from row boats, to sailing craft to a small steam launch. The San Jose boys constructed a two masted schooner at their sawmill. Running before the prevailing glacier breezes, the schooner profitably ferried prospectors and their supplies to the foot of the lake. Doc Ottawa sledded a 14 foot

Looking down on Sawmill Camp (Klutina Valley). A well stocked camp of goods all carried over the glacier.

Photo by Neal Benedict, Messer Collection, courtesy of the Cook Inlet Historical Society.

boat and small steam engine across the glacier reconstructing them on the lake. When the mass exodus began by mid-summer, he ran a prosperous business transporting four to six men at a time back up the lake against these winds. Throughout the summer, his shrill steam whistle could be heard echoing from the surrounding mountainsides.

From Peninsula Camp a trail led along the western edge of the lake to where St. Anne Creek (named after Anna Barrett who operated a restaurant on the lake) flows down from St. Anne Lake through a flat hay meadow known as Cranberry Marsh. About 80 tents were located here. This valley provided a low lying but swampy pass from Klutina Lake over into Tazlina Lake and was the route pursued by Captain West earlier in the season. Both Basil Austin and Dr. Townsend explored this alternate but difficult route to the Copper River. Will Crary while prospecting near St. Anne Lake found a white man's blazed trail, an old boat and a log cabin full of furs indicating that the '98ers were not the first whites to cross the glacier. Similarly, near the outlet of Lake Klutina Rafferty discovered a decaying clinker-built bateau and a rotten cache of clothing stashed in a spruce tree. Whether these items were left here many years before by Capt. West, Jack Sheppard, Peter Jackson or some other unknown early adventurer, we shall never know.

Jackson Delivers the Mail

Peter Jackson was to make his way many times through the camps along the glacier trail, down the Klutina and throughout the interior — this time not as a fur trader but as the mail carrier. Jackson on his mail route probably crossed the glacier and covered more miles in the interior than any other man that summer. In camps like Cranberry Marsh along the trail , the cry of "Jackson" suddenly breaking the silence meant "mail delivery" and people eagerly crowded around in hopes of hearing news from their loved one's at home and of the world at large. The big news was the

Looking West from Penin-sula Camp: "It was found to be an agreeable location for a camp because of its comparative freedom from mosquitoes, which probably were swept away by the cool breezes from the glaciers and were also discouraged by the fact that the peninsula had been burned over and nearly all its vegetation destroyed (Benedict, p. 69)."

Benedict photo, Carl Messer Collection, courtesy of the Cook Inlet Historical Society.

outbreak of the Spanish-American war. The importance of mail is brought home when one considers George Hazelet's diary entry stating that he intends to hike over the glacier to Valdez from Twelve-mile camp just to pick up and deliver his mail! "It only means a walk of 150 miles, nothing for this country" (Hazelet, 5/1/98).

Although the U.S. Government had established a post office at Orca, there was none in the more populous Valdez. Thus mail delivery to Valdez and along the trail was left to the whims of the Pacific Steam Whaling Co. and to private enterprise. According to Bourke, before leaving Valdez, the prospectors contracted with Jackson for a monthly mail pickup and delivery. The agreed upon price was $1 for each pickup and each delivery or supposedly $2 per month. Newspapers were delivered for 50¢. This arrangement worked fine as long as Jackson had a monopoly. However, others soon went into competition with him for this lucrative trade, and the Pacific Steam Company's agent in charge of mail in Valdez began distributing the mail among the carriers. Thus, rather than paying for a single delivery of mail arriving on a particular steamship, the prospector had to pay many times over. Leroy Townsend complains that mail delivery that season cost him $450. Furthermore, these new mail carriers soon earned enough for their passage home and abandoned the mail bags so that mail lay strewn along the trail and around Cordova, Valdez and aboard the various steam ships. Finally, when a post master was appointed for Valdez, a feud developed between the Orca and Valdez postmasters further delaying mail deliveries.

Schrader estimates that from August to October over 4200 letters were delivered to the Copper River area from Valdez. At $1 a delivery a mail carrier could really prosper in a short time. This is apparently what happened to Peter Jackson. A journalist for *Leslie's* who accompanied the prospectors on the trail writes:

"The lake was named Klutena, and the camp given the same name. Camp at lower end of lake consisted of many tents, with log foundations and a few log cabins in a row along the bankon left of lake. All faced lake and were a few feet back on a bench, 5 or 6 feet above waters edge. Main street was along fronts of camps. (Crary, 7/23/98)."

Photo by Neal Benedict, Messer Collection, Courtesy of the Cook Inlet Historical Society.

There is another Peter [besides guide Peter Cashman] of the glacier. He is Peter Jackson, born is Sweden, and for some years an Alaskan squaw-man employed in Salmon canneries. He became a mail carrier over the glacier and had soon earned $4,000. Prosperity was too much for him. He set about turning the money into whiskey, and the whiskey into himself. Other men have got the business, and Peter Jackson, if he outlives his spree, will be glad to go back to his meager wages of a salmon cannery. (*Leslie's,* 12/8/98)

By midsummer a large camp consisting of some 350 people and known as "Klutina City" had grown up at the outlet to the lake. The settlement even boasted a main street, "Mosquito Avenue," along which the majority of the 200 tents, shacks, log cabins, and winter caches faced the lake. On July 19th the little community elected a mayor and sheriff. The first item of official business was to pass a resolution stating that . . . "Abercrombie's statements regarding strikes and trails, as published in western papers, were false and unwarranted."

This justifiable animosity of the miners toward Abercrombie probably explains why many of his place names failed to stick. For example, in his official reports he calls Lake Klutina first "Lake Margaret" then "Lake Abercrombie," he refers to Valdez Glacier as "Bates Glacier" or "Bates Pass," and the Tonsina River as the "Archer River." None of these names are current today.

The Indians

A short distance downstream from Klutina City was an abandoned Indian village — the site of a summer and fall hunting camp. Most first encounters with the native population occurred here on the Lower Klutina. The Indians of the Copper River area referred to themselves as "Ahtna" or "people of the great river" i.e. the white man's Copper River. The suffix "na" in the local dialect means "river." They

What will be the fate of these poor Indians in the next two or three years it would be hard to forecast. Yet, while they are few in number. . . should they resent the wrongs that naturally follow from the contact of the white man with the Indian, the little band of Nicoli's following on the Chettyna could in one season, if so disposed, retard the development of this section of Alaska . . .(Glenn and Abercrombie, pp. 579-580). Photo by Benedict, Messer Collection, Courtesy of the Cook Inlet Historical Association.

numbered about 300 and were ruled over by two major chiefs — Chief Stickwan (variations Stickman and Stephen) to the north and Chief Nicolai to the south. The prospectors referred to them as "Stick" Indians or more often as "Siwash." Unlike the natives of Prince William Sound who were Eskimo, they were of Athabascan linguistic stock.

Before arriving, the prospectors were led to believe that they would encounter hostile Indians; however, nothing could have been further from the truth. In general, the natives were honest, friendly, helpful and hospitable. More than one starving and freezing prospector was rescued by natives. There were numerous instances of hospitality on both sides; Indians were often invited into prospectors' camps to partake of the white man's "muck-muck." The whites followed Indian trails and sometimes guides; often they relied on the natives for local geographical information and news of strikes and camps farther up the trail. Smart miners such as Hazelet even relied on the Indians for information on mineral locations. Hazelet follows Indian Charley's advice to prospect the upper Chistochina area, while Powell, who combined a little prospecting with his survey work, regrets his failure to listen to him.

Sympathy between the gold hunters and Indians seems in part to have been due to a curious coincidence of values regarding the sacredness of private property and a mutual abhorrence of thievery. Abercrombie writes:

> Each band keeps to its own territory while hunting and fishing, and resents any intrusion on the part of any neighboring band. It is not an uncommon thing, early in the season, for Indians on one side of the river to go hungry if the salmon running on the opposite side are on the territory of a neighbor.
>
> Their caches or balagans are never molested. When the spring thaw overtakes them, they cache their tobboggans or sleds without a thought of their

being molested by anyone. Their houses are treated with the same respect (Abercrombie, 1899, p. 578).

When so many of the gold seekers had overturned their boats losing their outfits in the lower Klutina, Abercrombie notes ". . . it was no uncommon sight to see these Indians wading out into the river and rescuing the supplies of some miner whose boat had been wrecked further upstream. Piling them up on the shore, they would go 3 or 4 miles out of their way to notify the owner where he could find them (Abercrombie, 1899, p. 578)."

Nonetheless, Abercrombie was not above the prejudices of his period. He hires Beyer's English speaking Indian wife, Omelia, as a guide.

> With her assistance I did fairly well in collecting data from the Indians and in impressing upon them our good intentions in opening up their country, until, on the occasion of some celebration, this squaw was for the first time in her life given some whisky, and, like some of her fairer sisters, being tempted she listened to the tempter and fell to rise no more. From that day forth I was unable to control her when there was whisky in camp (Abercrombie, 1899, p. 579.)

Here, he blames her weakness rather than the illegal dispensing of spirits to native Americans and the equally illegal importation of alcohol to Alaska.

Other commentators while admiring and sometimes pitying the natives on the one hand; with typical Anglo-American intolerance, view them as an unclean, weak and degenerate people lacking all civilized manners. This lack of tolerance for cultural differences is evident in the following humorous anecdote of Powell who fails to grasp the two cultures' differences relating to the treatment and status of women: "We met an Indian whose squaw and dogs were heavily packed. When asked why he did not carry the squaw's pack, he replied: 'Me got em dog to carry pack; squaw, he no got dog (Powell, p. 227).'"

The friendliness of the natives was in part because of their eagerness to trade for the white man's goods. After all, for the past century they had been making the arduous trip over the glacier or down the Copper River to the trading post at Nuchek; now the white man was bringing his goods to them. Favorite trade items were tobacco, tea and red bandannas; whereas the white men craved salmon, game and furs. During the latter part of the summer, many discouraged miners attempted to sell off their supplies at miners' auctions before leaving the country. With so many departing, the supply and demand market was soon flooded with goods, so that food and other coveted items sold at prices lower than Seattle and Valdez; firearms and clothing (other than footwear) were considered worthless. These were given to the natives helping to explain why so many appear in white man's garb in photographs of the period.

While the abandonment of food and clothing in the interior was a temporary boon for the natives, the longer term consequences of this fantastic invasion of

So many were going home. . . Clothing suffered the greatest cut, and for two months it would not bring over ten percent of its original cost. This was a "windfall" for the Indians, for they procured large quantities of clothing, and were often seen parading in the "white man's togs." (Margeson, pp. 135-136).

From Joseph Bourke Scrapbook, courtesy of the City of Valdez.

several thousand whites proved a disaster. Even with the population of 300-400 natives met by Lt. Allen in 1885, subsistence was difficult in this harsh, subarctic region. The paradise of the Klutina Lake hunting grounds was quickly destroyed by the invading whites. Thousands of trees were felled for firewood, boat, and cabin building. The formerly abundant game was soon slaughtered, driven off, or its habitat destroyed. Echos of the former buffalo days resound through the Lake Klutina diaries. "I found plenty of moose and bear in the country. Before the prospectors had driven them away there had been plenty of beavers (Lowe in Glenn and Abercrombie p. 591)." "Went fishing caught 84 salmon 45 trout at 7 mile camp (Pearson, 7/16/98)." "Oleson brothers. . . made a trip to Hudson Lake and bagged about two hundred ducks (Remington, p.54)."

The major source of devastation, however, was the numerous, habitat-destroying forest fires which raged through the dry Klutina valley that summer. Careless campers failing to properly extinguish fires were responsible for most of these. On the Lower Klutina, Joe Bourke and his men came close to losing their camp, boat and entire outfit to one of these blazes. Quick action and help from a nearby party in clearing trees in advance of the fire are all that saved them. Bourke remarks the amazement of some of his party returning from a scouting expedition down river.

> They were astonished at the wonderful change in the landscape during their short absence. Where they left a beautiful moss-covered earth and stately trees clothed in a dense growth of foliage, their return showed them nothing but the blackened ground, covered with ashes and acres of bare blackened poles. (Bourke, p. 55)

The wholesale destruction of the wildlife in this fragile area was to have a grave impact on the natives who by the winter of 1899 were starving and ravaged by tuberculosis and smallpox which the white man had brought with him into the

Few prospectors considered the impact of their activities on the land and wildlife resources or on the Ahtna Indians who were dependent on those resources for subsistence. In addition to forest fires started by careless campers, they over hunted and over fished. Conger, for example, reports catching 60 trout on August 20th, more on August 21st, 102 on the 22nd, and 70 on August 27th. (Holeski, pp. 128-129).

Photograph by Neal Benedict, Messer Collection, courtesy of the Cook Inlet Historical Society.

country. Augmenting the devastation occurring in the interior, the salmon canneries clogged the stream channels at the mouth of the Copper with fish traps allowing few salmon to reach the upper river. Thus the natives were further deprived of this major source for their winter food supply.

The Rapids of the Lower Klutina

The first four miles of river below Klutina City were fairly wide, smooth and only moderately swift but with a large dangerous rock in midstream. However, after this stretch, the river descends 600 feet in 25 miles to the Copper River in an almost continuous series of rapids, some worse than others. The current runs from 10 to 14 miles an hour. Townsend describes the river from here as a "miniature Niagara" and Bourke writes: "The upper (Klutina) river that we had just left was a pond compared to this . . . . The water in its ceaseless rush is tossed into seething foam over every rock and the river is full of them. A boat and its cargo to go through in safety must dodge every one of them or come to grief (Bourke, p. 48)." To which Rafferty adds:

> At this point the river descends in leaps and bounds over bars and bowlders [*sic*], with a deafening roar that has anything but a pleasant sound for a man who must risk his life and his precious outfit on its treacherous waters. Men who had faced the storms of the glacier for weeks, living on cold victuals, overcoming obstacles that would discourage any but the most determined, with never a thought of turning back, weakened at the rapids (Rafferty, p. 616).

Abercrombie with his usual tendency to exaggerate claims that 95% of those who attempting the river beyond this point wrecked losing their outfits. He sets a conservative estimate of the worth of lost outfits at $20,000. However, Bourke, Townsend and Guiteau would all disagree claiming that 30% of those attempting the lower river made it. Rafferty attributes this low success rate to poorly constructed

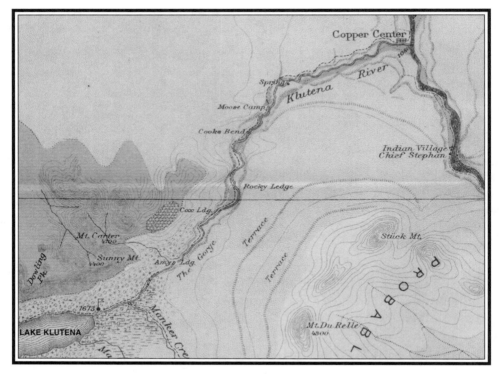

Lake Klutina to Copper Center. From the Geological Reconaissance map of a part of The Copper River and Adjacent Territory, Alaskan Military Expedition, 1898, in Glenn and Abercrombie, Explorations in Alaska, 1899. Courtesy of the Valdez Museum and Historical Archives.

boats and the inexperience of the river runners. Several experienced river men made $50 a trip running boats down the river for their owners.

Amy's Landing

At the large pool just before these rapids, a traffic jam of hesitant prospectors established a camp known as "Amy's Landing," named after W.S. Amy who had been one of the first '98ers to have camped here. Amy's Landing consisted of some 60 tents and about 300 people and became the major turn around point on the trail. Margeson notes that every day groups of twelve to twenty men with lightly loaded packs could be seen trudging back along the trail towards Valdez. He remarks that soon there were more "home seekers" than "gold seekers." (Margeson, 135).

Stories of the wrecks and goods strewn along the river downstream were certainly discouraging, but there were other alternatives to quitting. Some built cabins and caches, knocked down their boats to build sleds and waited for winter. The goods could easily be sledded down the frozen river after freeze-up. Others cached their goods temporarily and began laboriously hauling load after load on their backs down the river trail. Five hundred pounds could be advanced 1 mile each day. Still others used Amy's Landing as a cache and base camp for prospecting forays into the surrounding countryside.

Gold at Quartz Creek! The Stampede is on . . .

Just upstream of Amy's Landing a relatively easy pass led east up Manker Creek over a divide to Tonsina Lake. Gold seekers could prospect this area on short trips from the Landing. On August 15th, Fred Corles, who was "let out of" Treloar's party following a near knife fight just before reaching Klutina Lake, located the first placer gold claim at Quartz Creek near the outlet of the Tonsina Lake. By August 26th, fifteen prospectors had located claims — most of which were located on August 24th to 26th. Over the next two weeks, seventeen more claims were staked. When news of these strikes reached Amy's landing, they triggered a stampede. Margeson describes the ensuing stampede from Klutina City and Amy's Landing:

> Men left the camp at all hours of the night, and many started with but one blanket and very little provisions, seeming to forget everything in their eagerness for gold. It was told of one man that when he heard of this find, he just grabbed two biscuits, and ran. Another took with him but a single loaf of bread for a journey of ninety miles round trip. . . (Margeson, p. 206).

The first stampeders located claims on September 15th a month after the first discovery. In the following four days, fifty-five prospectors located claims. Claimants included R.F. McClellan who later participated in the great copper rush, several former members of the Connecticut Co., and Copper River Joe and his brother, Grant Remmington. Although Quartz and Manker Creeks produced some gold, the strike was a far cry from the bonanza most had been led to expect.

In fact, most of the Copper River Basin is unsuitable for placer gold prospecting. The geology of the area is against it. Although color can be found in most of the area's streams, to find paying quantities of gold requires digging down to bedrock where the heavier gold gradually sinks. However, almost the entire area is covered by massive layers of lake and glacial sediments which even the mighty rivers have not been able to penetrate in the last few thousand years.

By late summer, numerous prospectors were making their way back along the Klutina trail from various parts of the Copper River Basin and from the valleys surrounding Klutina Lake. Their stories were consistent: "the area was no good for gold." One of these, Neal Benedict, captures the disappointment:

> We awakened to the fact . . . that somebody had been laboring under a vast misapprehension; that the idea that gold lay about on the surface of Alaska and that any innocent who had the pluck to brave the glacier and get into the country beyond would become a Midas in the twinkling of an eye, was a very much mistaken one. . . . (Benedict, p. 110).

Lining down the Klutina

However, most of our diary writers were not among those who turned around at this point. Luther Guiteau, George Hazelet, Leroy Townsend and Horace Conger all successfully shot or lined their boats down the rapids while Charles Margeson, gaining access to a horse, packed his outfit down the trail.

Rather than risk their lives and goods in a single shot through the twenty five miles of rapids, many decided to line their boats down slowly or to shoot certain sections and line the difficult sections. To line a boat laden with supplies down the river, five men would attach a rope to the stern and brace themselves against the strong current while a sixth man aboard the boat would attempt to steer and fend off with a long pole. Because the men had to wade sections of the river where the banks were steep, the work was exhausting, cold, slippery and often dangerous. Sometimes only a mile would be made in a single day; all too often the boat would get away, go careening down the stream and be overturned or wrecked on a bar. "By this method there are more come to grief than by any other (Bourke, p. 49)."

Hell's Gate — Shooting the Lower Klutina

On June 8th, thirty bystanders, among them B. F Millard and Horace Conger, gathered at the head of Amy's Rapids to watch Luther Guiteau and Philo Snow make the first attempt at shooting the lower Klutina. Luther Guiteau describes the trip:

> It was a terrible, swift and thrilling ride. When we pulled out from shore at the head of the Rapids, we could hear all kinds of remarks from the twenty or thirty people who stood there to bid us goodbye. One fellow said, "We'll never see those suckers again." We came near believing him an hour or so later when our big boat shot up on top of a huge boulder quivered like a dying whale, but finally raced onward again after being struck by a tremendous wave.. . .
>
> We were both drenched to the skin by flying spray within twenty minutes of the start of the trip, and the roar of the white water was so great that we had to shout at each other to be heard. . . .
>
> A couple of fellows we had known on the trail yelled at us as we passed them. They told us to look out about a mile ahead when we came to a sign that said "Hell Gate"; and a mile still farther on, another that read "Devil's Elbow."
>
> Well, we saw them both, and will never forget them. When we picked up the big "Hell Gate" sign, our boat was going at least twenty miles an hour. We thought that we'd never make it through and if our craft hadn't been of absolutely the staunchest construction, it would never have stood the awful twisting and whirling and right-angle turns that it got as it was buffeted about in the raging waters that gave Hell Gate its name. . . .
>
> In spite of all predictions to the contrary, we shot the Rapids successfully, and did so in three hours and ten minutes. We were so completely exhausted when it was over, though, that we had to throw a rope to the Indians on the bank and be hauled ashore. (Guiteau, *Golden Eggs*, pp. 16,17.)

Copper Center

Those who managed to reach the confluence of the Klutina and Copper Rivers found a thriving community. One of the early arrivals, a shrewd Norwegian, Mr. Holman, laid out a townsite calling it "Copper Center." Here he set up tents for a hotel, store and unofficial post office later replacing them with log structures. There were 400 to 500 people and quite a number of cabins, tents and caches when Schrader passed through in late August.

Across the river from Copper Center, in June, Wisconsin gold seeker and surveyor, B. F. Millard, and a company of former Great Northern Railroad employees hacked a seventy mile trail north across the great bend in the Copper River creating a shorter path for the army's All-American route. Lowe followed this trail later in the summer so that on military maps it is marked "Lieut. Lowe's Trail;" however to the prospectors and in later tradition it is still known as the "Millard Trail." From Copper Center prospectors fanned out in various directions to search for the fabled Copper River gold.

However, gold rush participants like B.F. Millard, R. F. McClellan, Jas. McCarthy, Chas. Warner and mining engineer, Stephen Birch soon became aware of the country's copper potential and spent part of the winter of 1898-99 investigating rumors of rich copper deposits. Although they came into the country to search for gold, they found copper. By the spring of 1899, as we shall see, they were engaged in a major copper rush to find Chief Nicolai's deposits.

Besides its central location, Copper Center enjoys a commanding view of the spectacular, snow-shrouded Wrangell Mountains with Mts. Sanford, Drum, Blackburn and Wrangell, all over 12,000 feet, towering above the extensive, rolling

THe Millard Trail: "One of the most serviceable men we have is a Mr. B. F. Millard, a Surveyor by profession, but here in quest of gold. He cut a trail due north from Copper Center to the Tanana river, cutting of off the big bend in the Copper and lessoning the Journey by Seventy four miles (Bourke, p. 79)."

Photo by Benedict, from the Messer Collection, Courtesy of the Cook Inlet Historical Society.

"After supper I went out to take a look over the camp. Found it well populated with probably from a hundred to a hundred and fifty tents. One large one is used by Mr. Holman as a hotel and store but liquors are not for sale. Mr. Holman was one of the first to arrive in this section and laid out a townsite where we are now located on the left bank of the Klutina at its junction with the Copper (Bourke, p. 65)." Photo "Lower end, Copper Center," B62.1.1062. Courtesy of The Anchorage Museum of History and Art.

plateau of the river basin. The Mt. Wrangell volcano, active during this period, continually erupted steam and ash from its lofty summit. Copper Center was the major winter camp for the over 300 prospectors who refused to be discouraged.

Entertainment along the Trail.

Joe Bourke's diary describes a large Fourth of July celebration at Copper Center consisting of feasting, numerous contests and games, including, of course, the all American sport, baseball. These festivities were topped off with the local talent presenting a concert. Similar patriotic celebrations occurred all along the trail.

The prospectors brought their culture with them across the glacier. All sorts of musical instruments were packed into the interior that summer including cornets, guitars, banjos, mandolins, accordions, violins and harps. Both popular and classical music were performed. Along the trail, it was universally acknowledged that vocalist and piccolo player, Harry E. F. King, of the Margeson party was a top notch talent.

Although a heavy burden on the glacier, books provided prospectors many a long winter night's entertainment. Years later, F. W. Rosenthal recalls the importance of reading when weather restricted one's activities.

> There I was, all alone, confined to my cabin and nothing to read. I read the labels on the baking powder cans. I read the "Road to Wellville" in the grapenut packages. How glad I would have been to have Wm. Shakespeare or Chas. Dickens with me. Why I would not have objected to Bertha Clay or Mrs. Southworth. One day I wandered down to another deserted cabin and there I found a treasure — A book.
>
> I took it to my cabin. It was a book on fancy cooking. And there I sat eating bacon and beans and reading about lobster ala Newburg, pate de fois gras, fillet de boeff aux champs.
>
> I read that book a dozen times and then took a post-graduate course on

banquets that I found in the back of the book — Banquets of 20 courses, with different wines for each course.

Gentlemen, that was not a hardship. That was TORTURE. (Rosenthal, Presentation Speech, CS, p. 215).

Although books were at premium on the trail, undoubtedly the most popular form of entertainment was storytelling around the campfire. Powell relates how these storytelling sessions would sometimes be elevated into lying contests:

We were sitting by the campfire after supper, exchanging summer experiences with our visitors, when the Colonel told a very remarkable story. Whenever the Colonel tells anything, however, it *is* remarkable. He told us of once having discovered a wonderful deposit of lead on the summit of the Olympics. He peeled a flake of it, which he rolled down hill until it gained momentum by its weight, and then he lost control of it. He said it rolled down the mountain, eating deeper and gaining weight and speed until it tore up trees and left a great canyon as its track. There was silence in that camp for awhile, because no one felt competent to criticize the remarkable statement. Even Pete, our dog, had a doubtful expression on his countenance, but it was undesirable to reprimand him in the Colonel's presence. Cautiously remonstrating to the Colonel the next day, I said:

"Colonel, those visitors are strangers to us, and, while no one can dispute that remarkable occurrence, because you say you were alone at the time, they may be inexperienced in prospecting, and entertain doubts about it."

"See here," he replied, "if you don't sit right down on strangers at the beginning, they will impose on you. All young upstarts who come along invari–ably attempt to tell bigger lies than any one else, unless you knock them out at first, and then hold your club over them as long as they are in your camp. No, sir! I told *that* for self-protection, sir! It is a duty I owe to you and our camp, sir! We can't afford to allow ourselves to be imposed on, sir!" (Powell, p. 241 -242).

"After enjoying chicken stew for an evening meal, we were greatly amused by the Colonel's exaggerations. The Colonel was a most agreeable camp companion and very entertaining. . . . Some visitors came into camp while we were baking bread for our future needs, and the conversation turned upon cooking. One said he had cooked on a Yukon stove, another said he had cooked on a large hotel range, and the Colonel announced that he had cooked on a Cattle range. (Powell, pp. 228-230)."

Photo by Miles Brothers. Courtesy of the Valdez Museum and Historical Archives.

Chapter 6
Winter Exodus

Staying in the for Winter

By the fall of 1898, those still remaining had to make a choice: to return to civilization or to dig in for a winter. Photographer Neal Benedict and twenty-four others hired professional guide, Peter Cashman, to lead them back over the glacier in late August. The fall equinox storms soon made a return over the glacier extremely dangerous; so as winter approached, Abercrombie, Koehler, Powell, and Margeson all chose to float down the Copper River to Orca as did many others. Rafferty went north eventually making his escape via the Yukon River. Frank Schrader and his party descended the Copper to the Tasnuna and then proceeded over Marshall Pass reaching Valdez via the newly created military trail through Keystone Canyon. Austin, Hazelet, Guiteau, Conger, and Townsend all dug in for the winter.

No one knows the exact number who chose to remain in the interior that winter. Schrader's estimate of only 300 seems a little on the low side while Rice's later estimate of 600 is probably too high. (Schrader 368, Rice 97) The July *Seattle Post Intelligencer* estimated that 300 still remained in the interior following the mass exodus caused by the winter scurvy epidemic. Brady in his Governor's Report estimated that between 500 and 600 remained for the winter(*The Alaskan* 3/18/99). Quartermaster Brown records receiving reports from February 1 to April 9 of close to 300 scurvy victims or destitute people in the area. Some of these were probably present in Valdez, others still in the interior, and still others on the lower Copper River and at Orca. Those who remained in the upper Copper River region were spread out along the Klutina Valley from the foot of the Glacier to Copper Center, where most wintered. A small handful wintered at Hudson Lake (St. Anne's Lake). Active prospecting was pursued throughout the winter all along the upper Copper River from its headwaters on the Chistochina down to the Tiekel.

Winter in the interior began routinely enough with the men attending to finishing their cabins, gathering firewood, going on hunting trips; there was even time for a few concerts, poker games and a little local prospecting. However, as mid-winter approached and the mercury began to drop from 20° to 40° below zero, many began to experience a number of mysterious symptoms. When chopping wood, they would soon be out of breath and seemed to have little resistance to frost bite. The aches and pains in their joints, which they had attributed to the physical labor of hauling sleds over the glacier and boats up the rapids, seemed even worse now although their lives were much more sedentary. They noticed that they bruised easily and reddish-blue discolorations began to appear along blood vessels on the inside of their legs (blackleg); their gums bled, teeth loosened and jaws ached and swelled (big-jaw). Most thought they were suffering from rheumatism.

Scurvy, the Scourge of Prospectors

Scurvy, its symptoms, prevention and cure had been known to seafaring men since the 17th century but was a disease rarely encountered by the city dweller at the end of the 19th century; consequently most (but not all) were ignorant of the significance of their symptoms and of the necessary preventive measures. This ignorance was shared by a number of the prospector-physicians who consistently misdiagnosed the disease even though the first symptoms began to appear as early as April 1898 and the first death near the end of May.

Scurvy is a disease of the body's connective tissues caused by a lack of vitamin C which is usually derived from fresh fruits and vegetables. Cold, frostbitten extremeties, weakened by scurvy dramatically became swollen and gangrenous. Knowledge of the disease certainly existed along the trail as exhibited by Dr. Townsend's excellent essay appearing in Abercrombie's collection of government reports. Basil Austin mentions that he brought along a plentiful supply of dried fruit to prevent scurvy and notes that a man at Timber Camp cured himself by brewing tea from spruce needles which are rich in vitamin C. A few, especially those with a seafaring background, brought along vials of lime juice as a preventive measure. Luther Guiteau believes he protected himself by eating fresh moose meat rather than subsisting like most on a steady diet of beans and bacon. Certainly preventive foods lay close at hand as witnessed by the diet of the Indians who did not suffer from the disease. Most of the diary writers who escaped scurvy mentioned feasting on the abundant berries of the region while other urbanites for fear of poison berries avoided them. Copper River Joe, a pure food fanatic who would have felt right at home in any modern health food store, delivers numerous tirades on the evils of tobacco, alcohol, and refined sugar and flour. He rails against the poor diet of his fellow prospectors consisting mainly of refined flour, salted meat, bacon, and beans. His detailed knowledge of the causes and prevention of the disease illustrates that the necessary information was available to the common man at the time.

Wintering in Copper Center, French cook, Luther Guiteau, first became aware of the scurvy epidemic when Doctor Quick dropped by his cabin shortly after the new year to announce that he was treating two patients in a nearby cabin for either rheumatism or scurvy — he didn't know which. A few days later, Charlie Collins, arrived from up the Copper River with the intention of finding a cabin to establish a hospital in Copper Center. Eight or nine men up river were suffering from scurvy and frost bite and needed to be evacuated. A week later a cabin was donated by Doctor Nierman and on the 23rd of January the hospital opened its doors. Luther volunteered as cook, record keeper and male nurse. By this time, ten men from Copper Center had fallen ill with the disease and at least 20 men up river were suffering from frozen feet and hands because of scurvy.

Within two days of the opening of the hospital, one of the patients who was in very bad shape died. Despite the efforts of Drs. Quick, Nierman and Luther Guiteau's volunteer hospital staff working almost round the clock, two others soon followed. Quick and Nierman realized that they were beyond their depth, and they sent up the Klutina for Dr. Townsend. Arriving on January 31st, Townsend, who understood the disease, sprang into action. The twenty cases at the hospital were serious, but he was sure he could save them if the medicines held out. He began his treatment. However, anticipating still more cases, he knew he would need more medicine, antiscourbutics, fresh fruits and vegetables. He asked for volunteers to cross the glacier to Valdez to get supplies. When Brandon who had a good dog team and another volunteer left Copper Center on February 3, they carried the following letter from Dr. Townsend to Quartermaster Brown:

Dear Sir:

I beg to report to you herewith the serious condition which prevails in this camp. Scurvy has developed to an alarming extent. Two deaths have occurred and the hospital, which it was necessary to establish, is now full, and still throughout the camp are many who should be admitted. Up the Copper River the condition, I understand, is equally serious. Many are wholly without means and dependent upon others for nursing, medical attention, etc. We are in need, too, of some medicines, and any fresh fruit or vegetables which may be had at Valdez. Two have volunteered their services to bring in these supplies and are now on their way to Valdez. Several men are suffering with frozen feet, and amputations in a number of cases will, no doubt, be necessary. If it be within your power to give any assistance, I would respectfully ask your immediate consideration (Abercrombie, 1900, pp. 38-39).

Although Townsend's patients were responding well, he knew that he could not continue the treatments for long if Brandon did not return soon. However, conditions on the glacier were extremely treacherous this time of year and made travel slow and dangerous. Still Brandon did not return. Then one of those miracles that sometimes happen when things seem hopeless occurred: two prospectors from Ellensburg, Washington stopped by the hospital on their way out of the country and donated a quantity of medicine. Finally at the end of February, the volunteers returned from Valdez. As a result in late March, Townsend writes a letter full of hope:

The conditions existing. . .are greatly changed — marked improvement having taken place. Discharges from the hospital service have been so numerous that only 9 patients are now in my care. Six are leaving for Valdes today. The reports from up the Copper River are much more encouraging too. One death has occurred since I last wrote you, but I now feel justified in predicting no more, if we are at all fortunate in getting supplies in. Our vol– unteers, after a hard trip of 23 days, brought over 100# potatoes and about 40# of onions. To this we have since added 30# potatoes and onions and 6 dz.

Sledding on glare ice Copper River, Spring 1899: Despite the perilous crossing of the glacier, the mass exodus caused by the panic from the scurvy epidemic in the interior was expedited by the frozen rivers. Those that remained healthy used the frozen rivers to transport their goods. Photo No. B62.1.7 provided courtesy of the Anchorage Museum of History and Art.

lemons by purchase. This is the available supply at Valdes until the arrival of another boat.

Potatoes and onions sold at 50¢ per lb., lemons 25¢ each, etc. The camp now only numbers about 16, 9 of them sick (Townsend, Abercrombie, p. 75).

By February, news of the scurvy epidemic had spread throughout the interior camps. The panic that followed can be attributed in part to the mistaken notion that "the plague of scurvy" was contagious. Bryan Pearson estimates that by this time 75% of those along the trail were showing some signs of the disease. A mass exodus ensued; men decided that the known hazards of the glacier were preferable to waiting in the interior to be struck down by this dreaded "plague."

It was clear to the men in the interior that there would be no government rescue of scurvy victims that winter so they organized their own rescue missions. Those who were still healthy or only lightly affected began sledding the more acutely ill out of the country. Some like Varley and Cole would be taken out after break-up down the Copper River. This longer route, however, during the winter was quite hazardous. Townsend reports "Five or six of a party of 10 sick men who started from the Chittyna and points below for Orca were frozen on the way (Townsend, Letter, 4/13/1899)." Most were sledded over the glacier where a sudden winter storm could prove fatal. In November, at least two men had died from this cause. On November 14, nine men started over the glacier, two turned back and the rest continued on in spite of the raging storm. One man, Mike Smith, became exhausted, fell behind and was frozen on the glacier. The next day a party of at least five followed the above party and were caught in the same storm; they dug snow caves with their snowshoes and subsisted four days and nights on four quarts of beef stock and ten beef capsules. The storm blew itself out and they made their way down to Valdez; however, one of their party, Henry Krohn, died shortly after from the effects of exposure on the glacier.

The worst disaster occurred on February 26 when an entire party of six was lost on the glacier. Dr. Logan, who had been attending the sick at Twelve-mile camp on the upper Klutina, and three others were hired to sled two sick men out over the glacier to Valdez. According to Pete Cashman, Logan was a prudent man often warning others of the dangers of crossing the glacier in bad weather. However, on this occasion Logan apparently misjudged the weather; for on March 3 a relief party led by Pete Jackson and Melvin Dempsey found the frozen bodies of his party on the fourth bench. Dr. Logan's body was never recovered.

As spring approached, the patients at the hospital in Copper Center continued to respond to Dr. Townsend's treatment. It was becoming increasingly clear that help from the military relief effort was nowhere in sight. Dr. Townsend decided that the remaining five patients were well enough to travel; so he organized a relief effort of his own. He persuaded three Indians and three prospectors from Tiekel to join his hospital corps of four. Leaving on April 17, the relief party of ten made fairly good progress sledding the five sick men up the Klutina. By the fourth day, they were able to reach Sawmill Camp on the upper Klutina; however, from here the trail worsened. Townsend found a cabin at Sawmill for the sick. Leaving them in the care of his hospital staff, he and two of the Indians set out to cross the glacier to get the help of a larger party in Valdez. But storms raged over the glacier for the next three days as they waited in the tent hotel at its terminus. Here they were joined by a Mr. Knott who was also waiting for a break in the weather.

Finally on April 24, the weather cleared and the four were able to begin their traverse of the glacier. The recent storms had dumped two feet of new snow so snowshoeing up the steep incline was slow and tiring. They had hoped to reach the Christian Endeavor Society relief station on the third bench the first night; but the going was so slow that they were only able to reach the fourth bench where the bivouacked on the cold snow. The next morning when they reached the relief station, Townsend and one of the Indians were so exhausted from the exposure from the cold night spent on the glacier that they could not continue. Knott and the other Indian were able to make it on to Valdez for help. By this time Abercrombie and the Copper River Exploring Expedition had finally arrived in Valdez. One of Abercrombie's first actions was to send a relief party to bring the heroic Dr. Townsend down off the glacier. Arriving in Valdez on Wednesday, April 26, Townsend immediately set about arranging for the evacuation of the sick which he had left on the other side of the glacier. A military party was dispatched on Friday and his patients arrived safely on Saturday.

Abercrombies Report on Conditions in Valdez

Abercrombie's return to Valdez in the spring of 1889 is an oft-told story made even more sensational by the good captain's excellent prose style, flair for the dramatic, and tendency to exaggerate. Unfortunately, many historians subsequently writing about these events have treated his reports uncritically. It is no coincidence that these reports often read like well-crafted fiction.

On his arrival in Valdez, Abercrombie reports Quartermaster Brown as running out to greet him shouting "My God, Captain, it has been clear hell! I tell you the early days of Montana were not a marker to what I have gone through this winter! It was awful!" And indeed the scene in the hospital cabins as described by the Captain vividly captures this horror:

> I visited the various cabins in which he had housed some 80 or 100 of these destitute prospectors, and from what I saw there I was satisfied that while his remarks might have been forcible, they were not an exaggeration.
>
> Many of these people I had met and known the year before were so changed in their appearance with their long hair hanging down their shoulders and beards covering their entire face, that I do not think I recognized one of them. They were crowded together, from 15 to 20 in log cabins, 12 by 15, and in the center of which was a stove. On the floor of the cabin at night they would spread their blankets and lie down, packed like sardines in a box. Facilities for bathing there were none. Most of them were more or less afflicted with scurvy, while not a few of them had frostbitten hands, faces, and feet. Their footwear in some cases consisted of the tops of rubber boots that had been cut off by Brown and manufactured into shoes. Around their feet they had wound strips of gunny sacks, which were used in place of socks. Across the cabin from side to side were suspended ropes on which were hung various articles of apparel that had become wet in wallowing through the deep snow and had been hung up at night to dry. The odor emanating from these articles of clothing, the sore feet of those who were frozen, and the saliva and breath of those afflicted with scurvy gave forth a stench that was simply poisonous as well as sickening to a man in good health, and sure death to one in ill-health (Abercrombie, 1900, p. 15).

Abercrombie graphically described the horrible conditions in the military hospital in Valdez. However, photographs accompanying his report show a well kept hospital scene.

Photo from Abercrombie, 1900.

Left: Quartermaster Charley Brown.

Right: Capt. W. R. Abercrombie.
Photos from Abercrombie, 1900.

Undoubtedly, conditions were bad in the relief cabins in Valdez. Whether Quartermaster Brown, who was greatly respected by all, had done such a poor a job of caring for the scurvy victims as this description implies, we will never know. One must, however, be cautious in generalizing this description to the town of Valdez as a whole. To do so would be the equivalent of walking into the emergency room in a strange town then depicting these conditions as typical of that town. By the winter of '98-99 Valdez had a thriving business community consisting of at least three hotels, several restaurants, a pharmacy, a saloon, several outfitting and general merchandise stores, and the Christian Endeavor Society's Hall and Reading Room. Furthermore, minutes of the townsite meetings reveal a citizenry concerned with improving life in the new community and with helping the relief efforts on the glacier. The town had a active social life. On New Year's, the local talent organized a theatrical evening at the Roberts Hotel. Powell's description of the 100 men and nine women who passed that winter in Valdez contrasts markedly with Abercrombie's.

> This little colony represented a very small percentage of the four thousand people who had invaded that part of the then unknown. The task of overcoming the apparently insurmountable difficulties of exploring that great wonderland was left to them, with their indomitable will, energy and perseverance. . . . These pioneers would come out on the clear, crisp cold nights, and cluster in groups to witness the beautiful scenes that were enacted on the northern stage, where the sky-curtain trembled in dim aurora. We were embayed in calm seclusion in another world. . . (Powell, p. 92).

In further describing his visit to the hospital barracks, Abercrombie tells the story of meeting a large, raw-boned Swede with blazing eyes who tells him a story how a "glacier demon" attacked him and his son on the glacier.

> When halfway up the summit of the glacier, his son who was ahead of him hauling a sled while he was behind pushing, called to him, saying that the

demon had attacked him and had his arms around his neck. The father ran to the son's assistance, but, as he described it, his son being very strong, soon drove the demon away and they passed on their way up toward the summit of Valdez Glacier. The weather was very cold and the wind blowing very hard, so that it made traveling very difficult in passing over the ice between the huge crevasses through which it was necessary to pick their way to gain the summit. While in the thickest of these crevasses, the demon again appeared. He was said to be a small, heavy-built man and very active. He again sprang on the son's shoulders, this time with such a grasp that, although the father did all he could to release him, the demon finally strangled the son to death (Abercrombie, 1900. p. 16).

This is great storytelling and has captured the imagination of many. However, Abercrombie continues "When I heard this story there were some ten or twelve other men in the cabin and at that time it would not have been safe to dispute the theory of the existence of this demon on the Valdez Glacier, as every man in there firmly believed it to be a reality." He characteristically generalizes the men's superstition into one of his gross exaggerations that "70% of them were more or less deranged."

Other sections of Abercrombie's 1900 report reveal blatant exaggerations. For example, he writes of the panic in the interior caused by the scurvy epidemic:

To flee from these conditions was their one thought and topic of conversation, but where to, was the question on every lip, and when a number, regardless of the consequences, attempted what was considered an impossibility at that season of the year, the passage of the dreaded Valdez Glacier, leaving two-thirds of their party frozen to death on the vast ice fields, far up above the clouds, the panic was complete. If my memory serves me right, I do not think there was a single cabin in the Copper River Valley during the winter of 1898-99 that did not lose at least one of its party from being frozen to death or by scurvy (Abercrombie, 1900, p.19).

Clearly, two thirds (67%) of this "number" fleeing across the glacier did not freeze to death. No one knows the exact number of those who crossed the glacier that winter. The deaths were, however, recorded. Brown notes numerous winter crossings, many of them by large parties. For example, he cites one party of 25. If we calculate the percentage of freezing victims to those crossings from November through April based solely on numbers which appear in Brown's 1898 and 1899 reports, we get 7/49 =14% freezing victims rather than Abercrombie's 67%.

In preparing this book, we kept track of all the deaths on the trail from April 1898 to April 1899 recorded in our 22 original sources. Our count reveals: one died of accidental gunshot, one of a heart attack, one of blood poisoning, one from a bear attack, two from suicide, six from avalanches, five from drowning, seventeen from freezing (14 of these possibly on the glacier) and eighteen from scurvy for a total of fifty-one out of the original 3500. This count may be incomplete but is still better than a random guess. A conservative estimate of the number crossing the glacier that winter would be around 80. This would give us a figure 19% (14/80) dying from freezing — a figure much closer to Brown's than Abercrombie's. Abercrombie's assertion that not a cabin in the interior was without a death from either freezing or scurvy is equally absurd. If we use 400 as a median of the estimated 300 to 600 wintering over cited above, and if we assign an average of four men to a cabin we would have around 100 freezing and scurvy casualties rather than the 35 we have been able to document.

The general death rate appears to be approximately 50/3500= 1.4%. The death rate on the glacier would be 22/3500=.6%. Thus, even though the glacier trail was indeed dangerous, it was hardly the "death trail of '98" as it has so often been depicted. In fact, even after the trail up Keystone Canyon was established, the glacier route continued to be used as a shorter, more direct route to the interior. The *Valdez News Miner* records numerous crossings up to as late as 1905.

According to Abercrombie's report (confirmed by Brown) swift action was taken to offer relief to the destitute and those stricken by scurvy. In addition to ordering the rescue of Townsend's patients at Sawmill Camp, Abercrombie ordered Brown to rent additional cabins as a hospital, bunkhouse and mess hall for the sick and destitute and to equip them as best he could. He further instructed his clerk, John F. Rice, to cross the glacier and establish relief stations at Amy's Landing and Copper Center and "to extend relief and encouragement to the demoralized and destitute prospectors (Abercrombie, 1900. p. 20)." However, instead of scurvied miners, Rice found mostly the abandoned cabins of those that had fled the country the preceding summer, fall and winter. The prospectors' own relief effort had been a success. Once again the cavalry had arrived too late. The military relief effort was little more than a mopping up operation. Nonetheless, Abercrombie writes in his report that relief was extended to 480 persons. This number only becomes credible if it includes Brown's efforts at the end of the summer of '98 and during the winter and spring of 98-99. Powell reports "There were about 60 patients during the summer [of '99], fifty of whom recovered and the remainder were sent to their homes by the government (*The Alaskan*, 11/11/99)."

Although Abercrombie's belated relief effort was probably appreciated by most, Copper River Joe's account does not quite reflect the glowing report which Abercrombie submits to his superiors. He views the government's efforts as totally ineffective in relieving the suffering in the interior and as showing ingratitude for the prospectors who were the area's first trail-blazers. Furthermore, he criticizes surveyor Addison Powell who while working for the military asks to have some of the fresh sheep meat that he and his brother are packing back for the scurvy victims at Copper Center. In return Powell offers nothing for the scurvy victims from the fresh stores his pack train is carrying. Remington's criticism is echoed by prospectors A. D. Smith and R. V. Kirkham on their return to Seattle. They claimed ". . . the Captain is doing nothing in the way of rendering real assistance to the Copper River contingent. . . ." (Peggy Townsend cited from, *Seattle Post-Intelligencer,* 7/7/1899, Townsend Ms. p. 82).

Abercrombie, definitely not a believer in the welfare state, had a problem on his hands: how to arrange for the trip home for 80 to 100 destitute prospectors. His solution was ingenious. The prospectors needed money for the passage; he needed a road built, so he instituted a "workfare" program. Since he was authorized to hire men to build the road, he decided to employ 15 men at a time for one month at $1 a day (the going rate for comparable work in SE Alaska at the time was $3.25 a day) thus giving more miners an opportunity to earn their fare south.

However, the situation became complicated when Brown told him that some, who were spending their pay checks on alcohol smuggled in by a steamship company employee, were likely to remain destitute in Valdez. In perhaps the only case where the military intervened in civilian affairs in Valdez, Abercrombie, well aware of the political implications of interfering in the frontier liquor trade, instructed the employee to discontinue his deliveries and requested the company's shipping agent in San Francisco to stop the sale of liquor from ships calling in Port Valdez. Alas, never again would Luther Guiteau or other poor prospectors trapped on the glacier in a raging blizzard be visited in the middle of the night by an angel from heaven delivering a bottle of Canadian Club.

Furthermore, Abercrombie supplemented his "workfare" plan. Rather than pay the destitute prospectors for their labor, he allowed them to earn their passage. On the boat's departure, he gave them ticket and $5. The $5 was to insure that the prospector would not be arrested for violation of the vagrancy act when he arrived in Seattle. Abercrombie noted his program was such a success that "when one of the cutters of the Treasury Department . . . called at Port Valdez to take south destitute persons, it was found that none remained (Abercrombie, 1900, p. 20-21)."

Even with these stringent measures, the good republican Captain was still unable to protect the government from those who would attempt to benefit from its largesse. Brown reported that a Mr. Leunart, one of the party that Dr. Townsend brought across Valdez Glacier from Sawmill Camp had declared himself destitute. However, later Brown discovered that Leunart possessed certificates of deposit worth $2000. Appalled by this isolated incident, Abercrombie reports to his su–periors: "the majority of these men appeared to appreciate the spirit of the act under which the Department extended them relief from their distress. But there were many adventurers who, actuated by a sordid desire to save their money and live on the Government by simply misrepresenting their condition, deceived the agent in charge of the station by proclaiming themselves destitute (Abercrombie, 1900, p. 18)."

Brown and others remained optimistic that the hardships and disappointments of this first season had separated the true prospectors from the returning greenhorns and that the mineral wealth of the Copper River region would reveal itself to those who were willing to work during the next season. However, it was the returning discouraged and destitute greenhorns who would tell their tales of woe to the press and the outside world giving Valdez and the Copper River region a bad reputation. The December 8, 1898 edition of *Leslie's Weekly* carried the following headline and subhead:

Alaska's Awful Man-trap

The Valdez Glacier — A Grim Mountain Barrier between Prospectors and the Glittering Gold — Heights on Which the Elements Always War — An Ascent That Taxes Human Endurance and Courage — The Deadly Snow-bridged Crevasses — Abysmal Depths Where Dead Men Lie — The Descent into a Region of Summer.

Peggy Townsend notes the following article appearing in the evening edition of the July 7, 1899 *Seattle Times:*

> Tales of gold and stories of woe and hardship were brought down by passengers on the steamer *City of Topeka.*
>
> Miners from Dawson staggered down the gang-plank trembling under the weight of the gold sacks on their shoulders. Miners from Copper River staggered also, but not from the weight of gold. They were weakened by a terrible winter in the interior of a desolate country, that seems to be entirely barren of the yellow stuff that so many Dawsonites have found. (Townsend MS, p. 81).

Addison Powell relates with his usual incisive humor:

> Juneau had been the dumping ground for hundreds of stranded Copper Riverites, who had been shipped out at the expense of the government and steamship companies. They had given the Copper River country a bad name, and I astonished an interrogator by answering that I intended to return in the spring. I met him on the beach at the time, and when the astonished fellow recovered his speech, he called to his companion, who was some distance away, and said: '0 John! Here is a fellow from Copper River!'
>
> John replied that he had seen enough of those fellows, whereupon the first speaker answered:
>
> 'Yes, but this durn fool is going back!' (Powell, p. 100-101).

Chapter 7
A Glacier-free Route to
Alaska's Interior

Military Route and Road 1898-1899

One of Capt. Abercrombie's missions in 1898 and 1899 was to establish an All-American, glacier-free route across the coastal mountains to the Copper River Valley and beyond. Various political and economic factors combined to increase the importance of finding such a route. A few weeks, even a few days, crossing the Valdez Glacier with its crevasses, benches, avalanches, blizzards, whiteouts and high winds gave many a "natural prejudice ... against using any trail where it would be necessary to cross a glacier (Brown in Glenn and Abercrombie, p. 622)." However, the transitory phenomenon of a gold rush might not have provided a sufficient reason for investing public resources in locating and constructing an alternative route. Continuing disputes with the Canadian government over the boundary, taxes, and inspections added the necessary political motivation; the need for a less costly supply route to the interior provided the economic impetus.

The story of the search for a new, land-based route to the interior contains elements of adventure and danger, the excitement of exploration and the hardwork of trail blazing. During the summer of 1898, Abercrombie sent four scouting parties to search for a glacier-free route. Lowe, Brookfield, Cleve and Heiden each led parties that gained important new knowledge valuable to subsequent expeditions. By the end of the summer, Abercrombie's men had found a glacier-free route and constructed a primitive trail through the coastal mountain range. A fifth party, led by guide Pete Cashman, established the route from Thompson Pass to Quartz Creek, then over to the Klutina River, and finally down the Klutina to Copper Center.

Lt. P. G. Lowe and guide Harvey Robe began the military's search for an alternative route. On April 23rd 1898, Lowe and Robe headed up the large glacial river flowing into the head of Port Valdez in search of an old Russian trail rumored to lead to the interior. It was a futile search. If such a trail existed, rapidly growing alder bushes and rock slides now thoroughly concealed all traces. Abercrombie subsequently named the river in honor of Lowe, whose abilities and endurance he greatly respected. Addison Powell gives a slightly different account: "Lowe River was formerly known as Valdez River, but Lieutenant Lowe fell into it once and thereafter changed its name from its mouth to its source. According to this precedent, most of the rivers in that part of Alaska should be named Powell (Powell, pp. 80-81)."

On May 21st, Lieut. R. I. Brookfield, leader of the second party, left the cantonment to explore the so-called summer route over Corbin Glacier. This was Brookfield's second scouting trip. Three weeks earlier when the prospectors were confused by the discrepancies between the maps and reality, they had requested the aid of the Copper River Exploring Party. Abercrombie had placed Brookfield in command of the party that went over Valdez Glacier and confirmed that the prospectors were on the right trail. In a masterpiece of diplomacy, Brookfield writes in his Report:

> The only practicable winter route across the mountains then discovered led across the large glacier at the head of Port Valdez, Alaska, and this route had been taken by the 2,500 people who had attempted to reach the interior. Up to the middle of April the leading parties had not progressed beyond the summit of the glacier, and there was considerable doubt among them as to the route to be taken to reach the Copper River. It was with the view of quickly determining the proper trail and mapping the route that I undertook the enterprise under the direction of Captain Abercrombie. (Brookfield in Glenn and Abercrombie, p. 596).

Brookfield's reference here to the Valdez Glacier as a "winter route" seems to be an attempt to allow Abercrombie to save face regarding his faked 1884 glacier crossing. Brookfield writes that his second scouting trip was to explore the summer route:

> By way of introduction it should be stated that as it had been reported that there was a practicable summer route from Valdez to the Copper River over a glacier to the east, I was ordered to go over this glacier. This accomplished, I was expected to establish a camp on a lake which was said to be near the foot of the glacier on the other side (Brookfield in Glenn and Abercrombie, p. 593).

As the map on the cover shows, Corbin Glacier is an east-west flowing glacier like the one portrayed in Allen's 1885 map (see page 8). Brookfield's Report suggests that Abercrombie acknowledged that the north-south flowing Valdez Glacier was the correct "winter route," while maintaining that Corbin Glacier was the "summer route," that is, the route he took in the summer of 1884. Thus he charges Brookfield with crossing the Glacier and proceeding to the lake. If Abercrombie in 1884 attempted Corbin Glacier as a "summer route," we can now understand his arduous six hour climb to the terminus and its east-west trend. There is also less discrepancy between Abercrombie's reported 2,000 foot level pass and Corbin's 3,800 foot pass leading down Bear Creek than Valdez Glacier's 4,800 feet.

On Brookfield"s first attempt to cross Corbin Pass, his party encountered avalanche conditions. While he scouted ahead for a route, a snowslide killed some of the pack horses and the packers barely escaped with their lives. On his return, Brookfield found the packers had abandoned him. So caching his supplies, Brookfield returned to the cantonment for instructions.

By now, Abercrombie was convinced that the military expedition could not succeed without horses and departed for the states on May 24th, leaving Brookfield with instructions "to survey the trail as far as the summit of the Corbin Glacier, and if possible to find out and report upon the character of the country on the other side of the divide (Brookfield in Glenn and Abercrombie, p. 593)."

Lt. Brookfield's Expedition: Over the mountains and down the river

At ten in the morning on May 26th, Brookfield and a party of five left Valdez with light loads for a fast scouting trip over Corbin Glacier. Again, Brookfield encountered difficult snow conditions; but taking advantage of the nearly perpetual daylight, he reached his cached supplies seventeen hours later at 3 am. After breakfast, battling its way against a storm of rain and sleet, his party made the laborious, seven mile climb to the summit which Brookfield reports as 6,450 feet. This approsimates the height of East Peak. The pass is about 3,800 feet. Here, the sleet turned to snow obstructing the view. Since they were carrying three days' provisions in case of emergencies, Brookfield decided to continue until they found a campsite with wood and water. The party descended the northern face of the glacier until 8 pm when they reached its blunt terminus. There was no easy nor obvious way down off Corbin Glacier; and the mountains rose on either side in precipitous, unscaleable walls. Finally, wet and tired, trusting that the snow which had accumulated in the crevasses was strong enough to bear them, the men threw their packs ahead and slid down the steeper slopes. They would have to find another route back to Valdez. After sliding about 800 feet down the snow chute, they trudged down through Bear Creek gorge over and around large boulders on a route that Brookfield noted "was not practicable for any kind of travel." Eventually, they reached a wooded area. Brookfield reports "no lake was visible." Their relief at reaching a campsite, after 43 hours of nearly continuous hiking without sleep, was dampened by a disheartening discovery. In the valley below, the river ran south — not northeast into the Copper River. It was a bitter disappointment; they had not crossed the divide.

Brookfield was not even the first of the '98ers to reach the area. His party found prospectors already there who gave them additional food. They were, however, the first military men to reach the upper Lowe River area, which they surveyed discovering the long flats with canyons at both ends. Fleming and Dr. Lewis explored northward finding what they thought was the Kotsena valley. It was the Tsaina.

Although Brookfield could see a low pass beyond the Lowe River (Marshall Pass) that he thought led to the Copper River, he could not find a place to cross the Lowe which was a raging torrent. Whereupon he relinquished his desire to become the first to reach the Copper River by an alternate route and decided to return to Valdez by rafting down the Lowe River. This was not a simple feat. In addition to flooding, the river flowed through a narrow canyon whose precipitous walls plunged

This is the approximate view that Brookfield's party had of the Lowe River as it entered the canyon. In May of 1898, a large avalanche chute blocked their view of the river's course just before the far bend. The TransAlaska Military Trail, known locally as the "Goat Trail" which Lieut. Babcock and the prospectors built in 1899 appears as a dark streak above the cliffs on the right side of the river. Photography by the Miles Brothers, 1903, courtesy of the Valdez Museum.

hundreds of feet down to the water's edge. Steep cliffs and the base of a massive avalanche chute through which the river had scoured a course blocked their view downstream. They would have to run the river without first surveying the route for boulders, rapids, waterfalls and other avalanche chutes.

Brookfield describes their trip through the canyon:

> The raft being completed on June 3, a start was made down the river. The current was swifter than anticipated, and the raft was whirled and tossed about. Guiding poles were of no use in this swift current. The rapids were safely passed through, however, at the lower end of the valley, but once in the canyon the current would sweep the raft against the rocks, first one side then on the other, when the raft would frequently be almost overturned. The railing which had been constructed on two sides of the raft was torn off by the rocks almost at the beginning of the journey. The walls of the canyon were very precipitous, and on one side or the other they would generally be perpendicular to a height varying from 100 to 700 or 800 feet. At one place in the canyon the raft was stranded on a sand bar. Later on it struck a large submerged rock in midstream, where, under the strain of the rushing water and the combined weight upon it, the heavy cross logs snapped in two. Half of the raft held together, and after a time the men were successful in getting this portion of it off. Finally, however, the weight was too great, and, as the raft commenced to sink, Heiden jumped to the rocks as the logs were being swept near the right-hand side. Now the remaining portion of the raft turned bottom up, throwing everyone into the water but myself. Pope was carried away from the raft. He succeeded, however, in getting on the raft again, when almost immediately afterwards another rock was encountered, where the raft remained fast. Gardner had already jumped to the rocks on the left, and by his help the members of the expedition were able to get across safely to the rocky beach on that side. All of the men were wet to the skin and chilled with the ice-cold

water. We finally extricated ourselves from the canyon by a steep snowslide, which had come down to the water's edge at the lower end of the beach (Brookfield in Glenn and Abercrombie, p. 595).

Shipwrecked in the middle of the canyon, their clothes and bodies drenched in icy, 35 degree water, the men began a survival hike. Fortunately, they found and successfully climbed an 800 foot snowslide which led to a way over the mountain, down to the Lowe River.

On June 4th, twelve days after they had left Valdez with only three days supply of food, the party straggled back into Valdez. It took Babcock 23 hours without food or sleep from the time he started rafting down the river to reach Valdez. The men, exhausted and hypothermic, survived the canyon and numerous crossings of the Lowe River. However, according to Abercrombie, Brookfield never fully recovered. When he returned in July, Abercrombie writes that he found Brookfield "so much impaired from exposure and overwork" that he sent him back to the States.

Brookfield's expedition eliminated Corbin Pass as an alternative route, established that the area on the other side of the mountains was ice-free, reported the presence of two likely passes (Marshall and Thompson), traced the route of the Lowe River, and focused future efforts on finding a route through the canyon.

Today, thanks to the efforts of generations of highway crews who have periodically widened the canyon, hundreds of thousands of people drive through it. Similarly, hundreds of tourists, seeking adventure, raft the Lowe River with the aid of experienced guides and modern equipment perhaps never realizing the courage, foolhardiness and hardships of those who made the first trip through — Keystone Canyon.

At last, a glacier-free route, but is it any good?

On June 10th, while Abercrombie was still fetching horses, Hospital Steward, John Cleave, Corporal Heiden, and Private Studt made a third attempt to find an alternative route to the interior. Again, they traveled light, carrying a minimum supply of food and equipment. On June 25th, by keeping high on the ridge above Keystone Canyon they avoided all brush and reached the stream flowing from the eastern terminus of Corbin Glacier. Here on the flats, they met three Dutch prospectors, who gave them food and told them they had put a two-log bridge across the stream (Sheep Creek) in the next canyon.

Rain delayed their progress until they resolved to ignore it. Cleave observed that "the discomfort of being wet was of less moment than the task of carrying the additional weight caused by blankets, etc., being soaked (Cleave in Glenn and Abercrombie, p. 603)." They continued up the extensive flats north of Keystone Canyon, across the Lowe River and over a divide they hoped would lead to the Copper River.

To celebrate the Fourth of July, Cleave writes "we prepared ourselves a feast

Looking north up the Canyon: "This canyon is about 2 miles long. The walls on both sides are almost p e r p e n d i c u l a r, rendering it necessary, therefore, to find a route around one of its sides. I divided my party and sent men out both sides of the canyon to find the most practical route for a trail. (Heiden in Glenn and Abercrombie, p. 605)." Photo courtesy of Dorothy Clifton.

of duck, ptarmigan, beaver, and wild celery. While we were yet celebrating the sun came out. Unseen for many days, we welcomed it by immediately packing up and marching till long after midnight (Cleave in Glenn and Abercrombie, p. 603)." July 5th might have passed as just another miserable day bushwhacking to nowhere; but climbing over a knoll, they spotted the river they were following emptying into a much larger river. "This greatly encouraged us," observed Cleave. On July 7th, the three reached the Copper River just above its confluence with the Bremner. Filled with the momentary exhilaration of success, they debated descending the Copper River to Orca. However, because they were uncertain whether the Alaska Exploring Expedition would be recalled to fight on the Philippine front in the Spanish American War, they decided to return immediately to Valdez.

The continuing rains almost cost them their lives. When Cleave discovered that the streams were flooding, he hastily made a twenty hour forced hike to the glacier stream where he found "not even the rock on which the former bridge rested was now visible (Cleave p. 605)." Since the party was traveling light, no one had an ax for falling another tree across the torrent. They tried ascending to the stream's glacial headwaters, but a shortage of food and nearly impassable terrain forced them to return to the former crossing. Finally, a party of six civilians with axes and ropes rescued them. Once again, the Dutchmen fed the nearly starved soldiers.

Cleave's initial euphoria faded with the realization that although they were the first to cross Marshall Pass, it was not a good route. It reached the Copper River below the nearly impassable Wood Canyon. The best route needed to connect Valdez to the Copper River above Wood Canyon near Copper Center.

The military was not alone in seeking a better route to the interior. One of the old-timers, mail carrier Pete Jackson, found a route suitable for horses at least by June 28th 1898. Crary reports that he arrived at Lake Klutina with the mail and two pack horses via an eastern trail "about half way up ridge, said to lead to Valdez (Crary, Diary, 6/28/98)." After crossing Keystone and Thompson passes, Jackson probably pioneered the route over the Quartz Creek divide down to Lake Tonsina and from their to the Klutina River. Unfortunately, neither the mailmen nor prospectors who pioneered these routes, left records.

Corporal Heiden wins the Day

While Abercrombie was making his pack-train traverse of the glacier into the interior, the search continued for an alternative glacier-free route suitable for pack animals. On September 2nd, Corporal Heiden, who had been on both Brookfield's and Cleave's expeditions, departed Valdez for Keystone Canyon with a party of three enlisted men and ten pack animals. He, too, encountered heavy rains that "caused the glacial streams to swell " washing the men and horses off their feet several times and making the Lowe River impassable. They spent the time waiting for the rain to stop "cutting trail through the dense growth of alders and graded the hillside, which was extremely hard work (Heiden in Glenn and Abercrombie, p. 605)." When the rain and river subsided, Heiden moved camp to Keystone Canyon whose "walls on both sides are almost perpendicular, rendering it necessary, therefore, to find a route around one of its sides." The route through this section was so difficult that they had to make three camps in 2-1/2 miles, but once on the other side, they "had about 4 miles of comparatively easy traveling (Heiden, p. 605)," through the area they named "Dutch Flats" after the three Dutchmen who befriended Cleaves' party.

At the head of Dutch Flats, Heiden chose the northern route, possibly Jackson's, that everyone hoped would connect to the claims at Quartz Creek and to the Copper River above Woods Canyon. He expected to meet Schrader who was instructed to pioneer the other end of the route from Copper Center over the Tsaina. As the rain turned to snow, Heiden's men cut a trail toward the pass through miles and miles of alder brush and windfalls. Meanwhile, an advance party returned with the welcome news that a pack train could make it over the pass.

The mission was a success: they had cleared a trail through the Keystone area and confirmed a pass to the Tsaina River on the other side. Abercrombie named the pass "Thomson Pass" after the "Honorable Frank Thomson of Pennsylvania." Years of mispronunciation corrupted the name to "Thompson Pass."

Five weeks had passed. A messenger arrived with the discouraging news that Schrader's party had not been able to find a route up the Tsaina and had gone up the Tasnuna instead. Heiden returned to the Tasnuna [Marshall Pass] and "After three days' travel . . . found them, without horses and almost without provisions, and learned from them that they had lost their whole outfit and one man was drowned in trying to cross the Archer River [Tonsina] on a raft (Heiden, p. 606)." Schrader, after meeting Heiden's detachment, returned to Valdez by Heiden's trail.

When Heiden stopped work for the season, he had made a trail through the Keystone Pass Area that was "free from glaciers" so "the traveler does not suffer from want of wood (Heiden, p. 606)." The route would be "available all year round to the Copper River Valley" and "with but small difficulties to overcome, a railroad could be built to the Copper River country by this route. . .(Heiden, p. 606)."

Pete Cashman, King of the Mountain Guides

On October 18th, guide and packer, Pete Cashman, with Charles Anderson, Jack Stewart and Joe Ham set out on foot over Heiden's trail to collect 13 horses left by Abercrombie's party with Chief Nicolai at Taral on the Copper River. From Valdez, they went to the end of Heiden's trail, then explored several valleys before locating the pass leading to Quartz Creek down the Klutina River to the Copper River finally arriving at Taral below Copper Center.

The Indians at Taral knew that the return trek would not be an easy one at this time of year. When Cashman tried to repay their hospitality, they refused his gift:

> They gave us to understand that if we were short of grub to come back to them and they would supply us. They said; "White man ha-low muck-a-muck. Indian high-you muck-a-muck. One moon high-you cold white man no muck-a-muck. Indian pot latch hi-you muck-a-muck. In one moon high-you cold, high-you wind, white man die," which we found pretty near right. (Cashman, p. 164).

Cashman tried to take the horses out via the Copper River, but turned back after ten days when the rugged terrain around Woods Canyon slowed his progress, and it was clear that he did not have sufficient supplies. He headed back to Copper Center in bitterly cold weather with little or no food. When Cashman wrote his report, he seems still to have been hungry.

> We arrived opposite Copper Center at 4 o'clock Thanksgiving Eve. . . When we got to Mr. Amy's cabin they could not believe it was us, as they told us it was over 65° below zero the last three nights when we stood around the camp fire. We were so hungry we ate supper at Mr. Amy's cabin, then went to Mr. Fisher's cabin and had another good supper. We still felt empty, so we went to the hotel and had another supper. Notwithstanding this we still felt hungry (Cashman in Abercrombie 1900, p. 166).

Cashman eventually made it back to Valdez via the glacier in February, reporting that a good trail to the Copper River via the Lowe was feasible.

Congress approves funding for an All-American Route

Abercrombie and the prospectors both desired a road to the interior. Commander Abercrombie exercised considerable power in recommending a route, in setting the construction schedule, and in arguing for its economic feasibility. Although the miners prided themselves in blazing the first trails, they believed nonetheless it was the government's responsibility to build roads.

In his 1898 report, Abercrombie notes that when the Expedition met returning miners on the trail, "they cursed . . . the commander of the 'Government outfit,' for not sending men into the interior ahead of them to cut trails and mark out routes over which they might travel to prospect the country (Glenn and Abercrombie, p. 588)." In the end, however, Abercrombie concludes with the recommendation that the government build a military trail "through Keystone Pass . . . from the head of Port Valdez, Prince William Sound, to Thomson Pass, at the head of the Tonsena Valley, a distance of about 35 miles. . . (Glenn and Abercrombie, 1899, p. 589)."

Abercrombie did not request funding for a road to Eagle City. Congress, however, in 1899 appropriated expenditures to "open up a military road to Copper Center, and from [there] . . . the most direct and practicable route to Eagle City (Abercrombie, 1900, p. 9)."

Starting the Trans-Alaska Military Trail

Before Abercrombie returned to Valdez in April of 1899 to begin work on the Trans-Alaska Military Trail, prospectors like Millard, Amy, and McClellan were already extending Heiden's trail. They had heard stories of copper "float" (rocks plucked from their bedrock and carried downstream), heard of Young and Hoffman's copper claim, and heard stories of nuggets of almost pure copper that a prospector had purchased from Chief Nicolai. In the great race to locate Chief Nicoli's copper deposit, McCarthy, Millard and others did not wait for the soldiers to build them a road. They did it themselves. However, prospectors cutting trails generally consid-

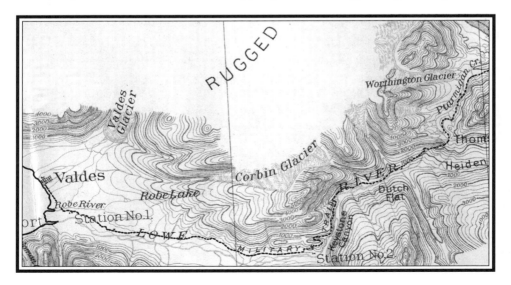

ered only the easiest and fastest route.

By contrast, Lieut. Walter C. Babcock who mapped and organized the 1899 road building efforts had to find a route that could withstand heavy pack-train traffic, year after year. He began work on April 26th from Station No. 2 near Keystone Canyon. In siting the route, Babcock had to consider the gradient, which had to be suitable for pack trains and dog sleds; the number of river crossings, the need for bridges and the risk of flooding. He also had to take into account the hardness of the terrain; horses' hooves could turn soft or boggy areas into a muddy quagmire.

Babcock divided his work force into three components: the advance crew surveyed the terrain and laid out the course; advance workmen brushed out the trail; and the trail gang actually built the road. Babcock began the road about 1/2 mile west of Keystone Canyon, switchbacking it up to a bench about 700 feet above the river, then sloping it down to a natural bench about 300 to 450 feet above the Canyon.

He soon appreciated the efforts of his advance team as they struggled along the steep slopes covered with slide alder:

> The effort of climbing over, under, and through this brush on a side hill so steep as to scarcely afford a foothold; falling, stumbling, grasping at the devil clubs; bruised and beaten by the stout alder branches, and, at the same time, endeavoring to blaze out a line with a uniform grade or on a level is simply inconceivable to one who has not tried it. Frequently this has to be done many times before a line is secured that is considered the best possible (Babcock in Abercrombie, p. 61).

It took the advance party six hours to cover the first three mile section of Keystone Canyon and more time was required to return to camp by a slightly different route. They did this four times before selecting the final route. Conditions were no easier for the work parties. Axmen "had to support themselves by hanging to the brush with one hand while they chopped with the other (Babcock, p. 61)." To make a five foot roadbed, graders had to made 15 to 20 foot vertical cuts. The four miles through Keystone Canyon crossed fifteen mountain streams, eleven of which required the construction of retaining walls. One bridge was built.

Not only was the Keystone Canyon section the most difficult, it also had to be completed in a rush so the post-office inspector and his outfit could begin their trip to Eagle. On June 11th, Abercrombie dispatched 11 more men to help and "by 5 p.m. Saturday, June 17th, the trail was ready for the pack trains to go in." Babcock writes with obvious pride that "the key of the whole route, was constructed in thirty-five working days (Babcock, p. 62)."

Even though the route beyond Keystone Canyon to Copper Center still was not finished, government parties and prospectors started moving over the road. On June 18th, John Rice accompanied Mr. Wayland, the postal-inspector, and five pack horses to test the feasibility of a mail route to Eagle City on the Yukon. On June 19th, 38 more pack horses and 9 dogs went north; and so it continued all summer. A

Post Office Inspector Wayland's party. "Post Office Inspector C. L. Wayland accompanied the expedi–tion for the purpose of establishing post-offices at the several mining camps along the line of travel. Our course was up Lowe River to Keystone Canyon. . . In the matter of picturesque scenery the Key–stone Canyon is one of the finest in Alaska. We passed through the canyon and down into Dutch Valley June 19." (Rice in Abercrombie, 1900, p. 95). Photograph from the Bourke Scrapbook, courtesy the City of Valdez.

government pack train and the "copper-rushing" Young and Downy party, raced along the new trail to catch up with the combined Millard, McCarthy and McClellan parties. By July 27th when the trail was completed to the summit of Thompson Pass, traffic was so heavy Babcock had to dispatch a man with sufficient tools to spend the rest of the season patrolling and repairing the road. Babcock writes to his superiors that "Many prospectors were constantly coming and going over the road, and all expressed their satisfaction and relief at having a road to travel that avoided the dangerous Valdez Glacier, and shortened the journey from the interior by several days (Babcock, p. 64)."

In mid-July, Babcock decided that he, Mr. Worthington, his transit man, and two packers would be the advance party to map the route from Thompson Pass to Copper Center. From Thompson Pass, they followed Ptarmigan Creek down to Tsaina then along the edge of Tsaina Valley, across the low divide to the north fork of the Tiekell River, where they worked their way through the remains of a recent forest fire, past Stewart, Boulder, Ernestine and Falls Creeks. Here they reported abandoned prospecting camps and a few prospectors still at work. They then continued across the low divide to the Tonsina Valley.

Babcock and Worthington found the stretch between Stewart and Boulder Creeks the most difficult to map because the stream, wandering extensively back and forth across the valley, had undercut banks and deep pools. Many beaver dams pocketed the valley, but the prospectors had killed off all the beaver. Heavily forested, the valley was covered with a thick understory of alders and willows, currents and wild roses — all seemingly designed to obstruct a surveyor's view. It took them five days to map the seven miles to Boulder Creek where they encountered the charred remains of Tiekell City.

At Tonsina, with the summer running out, Babcock encountered wet, boggy ground totally unsuited for packhorse travel. Discouraged, he and Worthington

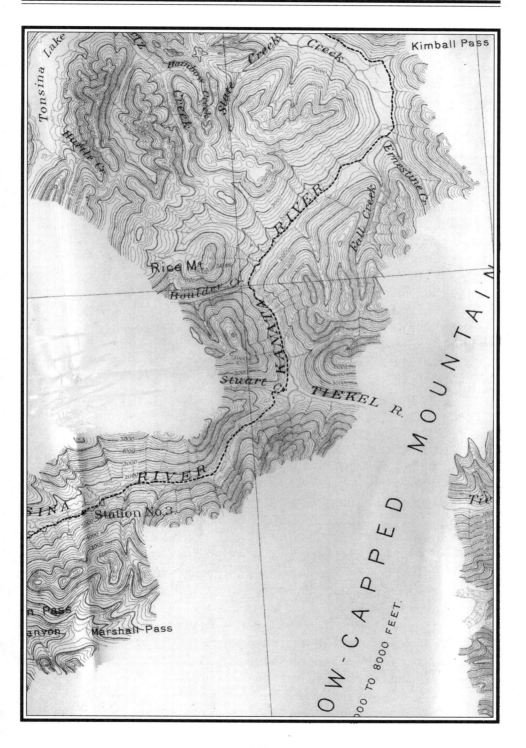

criss-crossed the area looking for a route. Luck was with him, however. His new packer, Christopher Tjosvig, arrived. Tjosvig, a prospector, had explored the Quartz Creek and Tonsina area in 1898. He showed Babcock a more direct route to Tonsina that ran through the divide between the South Fork of the Tonsina and Bernard Creek. Babcock named this Kimball Pass. At Bernard Creek, he sketched in the military road across the Tonsina and then along Trout Creek and over the last bit of coastal mountains down to a place on the Klutina about 8 miles above its confluence with the Copper River. A good bridge site here had already been selected. Babcock concluded the summer feeling satisfied with both the geology and grades for the final stretch. The military could build a road to Copper Center.

By the end of the season, Babcock's road gangs had completed 93 miles of the Trans-Alaska Military Trail which were suitable for packhorses, 35 miles which were in excavation, 67 miles cleared and brushed, and 12 miles cleared only. In addition, they had built 26 bridges, not including culverts, totaling 856 feet. The width of the road varied from five feet in Keystone canyon to ten feet elsewhere.

The Richardson Highway now follows approximately this route. In many cases, the roadbed has been shifted to the valley to avoid avalanche slopes. With the aid of modern road building equipment, the route now lies on a gravel roadbed across the wetlands — an option not available to Babcock.

When Rice returned from his trip to Eagle City with Postal Inspector Wayland, he confirmed that a road could be built all the way to the Yukon.

> The route over which we traveled from Port Valdez to Eagle City presents no such obstacles as the route through the Cascade or Rocky Mountains. The pack trail now constructed through the Coast Range by Captain Abercrombie could be transformed into a wagon or railroad bed. No glaciers are to be encountered nor any other serious obstacles. (Rice, in Glenn and Abercrombie, p. 102).

At the end of 1899, the Copper River Exploring Expedition had established the feasibility of an All-American, ice-free route to interior Alaska and the Klondike Gold Fields.

However, the summer's survey work was not without its mishaps. In crossing one of the glacial streams, which Babcock describes as only 15 feet wide and about 18 inches deep, the expedition almost met failure. Babcock reports that just as Mr. Worthington reached the far side, he was struck by a moving boulder. Suddenly, he disappeared in the water and was swept downstream, while trying to hold onto his gun and the surveyor's transit. Babcock ran alongside. "After a few minutes' struggle he crawled out on the side from which he had started. He still had hold of the gun, but the transit and hand ax were gone (Babcock, p. 65)."

The situation was serious. Without a transit, they could not complete their summer's assignment. Since it was late in the day, when the stream, fed by meltwater was at its highest, Babcock and Worthington camped waiting until morning when the stream flow would be less. They could then try diverting the stream. The next morning, to their relief, they found the transit caught in a snag a quarter of a mile downstream, damaged, but fortunately still usable. The glacier near Worthington's mishap today bears his name.

Prospecting results in 1899

In surveying the route for the road, Babcock collected information on mining prospects noting most claims appeared abandoned, except for those at Tiekell, Falls and Ernestine Creeks. At Tiekell, Babcock encountered a few prospectors who were finding fine gold flour but "nowhere sufficiently plentiful to pay for the labor of working a claim (Babcock, p. 70)." However, prospectors at both Fall and Ernestine Creeks were extracting gold in paying quantities.

At Boulder Creek, Babcock made a side trip to Quartz Creek, the scene of the most promising gold strikes in the previous fall and winter. On August 16th, he reported reaching the camp of "Mr. Amman and his wife, and here I saw the first signs of serious gold placer mining. They had as yet found no gold in paying quantities, but both were patiently digging away in the hope of some day reaching bedrock (Babcock, p. 70)." On returning to the military trail, he camped near the Manhattan Mining Company's abandoned tents noting "the owners had either left the country or were at work in the Government trail gang, earning sufficient money to pay their passage back to the States (Babcock, p. 73)."

Babcock's reports reveal, somewhat sadly, that even those who stayed and worked hard, very hard, were not finding the promised gold. Gone from his writing are the optimistic comments, like quartermaster Brown's, that once the green horns and slackers left, the "true prospectors," hard working men and women, would find

Slate Creek. Eleanor Doty and Anna Barrett. Photo B62.1.1079. Reprinted courtesy of The Anchorage Museum of History and Art.

Melvin Dempsey, standing in the Congregational Church entrance, was President of the Christian Endeavor Society from 1898 until 1903. He is credited with being the major contributor to the construction of Valdez's first church. At the encouragement of Reverend Cram, Mrs. Brownell replaced Dempsey as President on May 12, 1903. Dempsey resigned from the church on January 5, 1904. Photo

the promised gold. As the summer of 1899 drew to a close, the awful truth was becoming apparent — neither gamblers nor hard workers could find gold when there was none. For the remaining few, who believed that hard work led to success or that God rewarded those who labored, it was difficult to quit.

Other mining districts showed more promise. Both Dempsey and Hazelet found placer gold in the Chisna and Chistochina districts — the area Powell identifies as West's gold strike. A mining district was established with Dempsey as recorder. Powell, who visited the site while surveying for a shorter route to the Lake Mentasta area, reported it would take another season to determine "the extent of pay dirt and its value (Powell, 1900. p. 133-4)." Other prospectors, more confident, "jumped" Hazelet's claims over the winter. Later, Hazelet would blame Dempsey for this.

Although miners' meetings could resolve many issues, claim jumping required lawyers and courts. Disputing claimants spent the 1901 season filing court papers against each other. Hazelet's ownership was finally resolved in the spring of 1902, and serious work begun. Unfortunately, it did not save Hazelet from bankruptcy. Dempsey, however, earned enough from his claim to contribute significantly to the construction of the Congregational Church in Valdez and to build himself a "mansion" near his claim.

In the end, the Copper River Basin produced very little gold for such an exten–sive area. The only really significant finds were on its northern extremity in small area at the headwaters of the Chistochina comprising Slate Creek, Millers Gulch, and the Chisna. These are all located on the southern slopes of the Alaska Range and are a part of the same mineral province as the gold-rich Tanana just across the range on its north side. Small strikes were also recorded at Quartz Creek, Tiekel, Ernestine and Fall Creeks and on the Nizina and Bremner Rivers. To put the significance of these discoveries in perspective, by 1905 these areas collectively (including the Chistochina) had produced less than one and a half million dollars in gold. The Ramsey-

After the discovery of Chief Nicolai's copper deposits, Millard organized a party to survey for a railway route from Valdez to McCarthy. The Miles brothers accompanied the survey party as phtographers. Here the brothers and two other members of the party stand on a large piece of copper "float." Photo courtesy of the Valdez Museum and Historical Archives.

Rutherford, Cliff, and Mineral Creek hard rock mines and the Midas and Ellamar copper mines, all on Prince William Sound within 25 miles of Valdez, boasted a total gold production of over two million dollars. It is an irony of history, that to find gold in 1898, the stampeders need not have endured all of the toils and hardships of traversing the glacier, descending the Klutina and ascending the Copper River.

The big strike, the strike that created a boom in Valdez, the new town of Cordova, near Orca, and a railroad to the new town of McCarthy came not in gold but in copper. The combined parties of McCarthy, Millard, McClellan, Warner, Amy, Fitch and others beat Downy and Young to Chief Nicoli's copper deposit. They found it not by searching the mountainsides but by enlisting the services of the old Chief Nicolai himself. In exchange for all the supplies left by the Allis party on the Copper River the previous summer, Chief Nicolai, who was ill and whose people were facing starvation, provided a guide who led the prospectors to one of the world's largest copper deposits, the Nicolai Mine near McCarthy. Although B.F. Millard, Steven Birch, R. F. McClellan came into the country to search for gold, they found copper. And in the end, the corollary copper rush proved to be the more significant event for the development of Valdez, Cordova and the Copper River Valley.

A road or railroad to the interior was now a necessity.

Finding a Railway Route

As if anticipating the huge copper find, Abercrombie's party included railroad engineer, Edward Gillette, who spent part of the summer of 1899 surveying a route. His initial reaction was somewhat less than optimistic:

> The stupendous masses of mountains and ice-filled canyons and valleys
> back of the green wooded islands along the seacoast, while forming probably
> the grandest scenery on this continent, gives no encouragement to the explorer

or engineer in search of a practicable route for a railroad into the interior of the country, combined with that of starting from a good harbor. (Gillette, p. 146)

After making his survey, Gillette decided Valdez had a good harbor and a railroad route better than that of Skagway's White Pass and Yukon Railroad:

> It is of considerable value to this country in having this main route for transportation within its own territory and, consequently, jurisdiction. Some of the many complications which have arisen in the Canadian Northwest Territory will be eliminated, and Alaska developed without the hindrance or consent of a foreign country. This deserves our patriotic consideration. Our prospectors will have an opportunity of getting into the region at the head of the Tanana River and its eastern tributaries, and on soil belonging to the United States. (Gillette, pp. 147-148)

Abercrombie supports Gillette's conclusions stating "The future for a railroad through this section is, in my opinion, very promising, owing to the presence of large zones of heavily mineralized copper deposits, the development of which will unquestionably yield a local tonnage of great volume (Abercrombie, 1900, p. 28)." Millard, McClellan, Amy and the other nineteen prospectors who shared the Nicolai Copper Claims knew what they were doing when they gave Abercrombie a 1/23rd share in one of the richest copper mines in the world.

After a year of news stories describing the Valdez Glacier route as a "Death Trail," the townspeople must have been pleased to hear Gillette praising Valdez as a railroad terminus and predicting "Valdez may well lay claim to being the main gateway for Alaska Commerce." When the government published Gillette's report in 1900, it inaugurated a long-simmering debate: should railroads be built with public or private capital. Proponents of public ownership argued that because the rates would be equitable, more mines could be developed and the price of minerals

Freighting near the top of Thompson Pass on the Trans–Alaska Military Trail, which became the Valdez-Fairbanks Trail. Following the discovery of gold in the Fairbanks area, Valdez enjoyed the prosperity and the problems of other gold rush towns. Photo B80.41.44 by P.S. Hunt from The Anchorage Museum of

lower, which would be a benefit to consumers; others argued that capitalists should make the investments, make the decisions on rate structures, and make the profits.

The Government Reports also inaugurated the Valdez Railroad Rush of 1901 to 1907. For example, The *Valdez News*, which began publication on March 3, 1901, carried fourteen front page stories concerning various railroad schemes for Valdez during its first year. John Healey visited Valdez promoting his Valdez to Siberia Railroad and Heney arrived in Valdez to survey a route to Eagle City; owners of the Burlington & Missouri Railroad, Chicago, Milwaukee & St. Paul Line, and the English Close Brothers, builders of the White Pass & Yukon Railway were all said to be interested. Finally, in October 1901, the Burlington Company obtained the right of way through Keystone Canyon.

Railway fever continued until 1907, when the Reynolds backed Valdez "Home Railway" declared bankruptcy following a shoot-out with a competing railroad company in Keystone Canyon. The ensuing scandal brought down Governor Brady, the District's much respected governor, who then spent two years trying to help the shareholders recover their losses. The Guggenheim-Morgan Syndicate built their railroad from the new town of Cordova. For almost three-quarters of a century, the loss of the railroad haunted Valdez. When oil companies chose Valdez as the terminus for the TransAlaska Pipeline, the City and its citizens worked hard to assure that this time Valdez would be the site finally selected.

Ft. Liscum

Abercrombie's 1899 mission included establishing military reservations. After he and his men fought the first of many floods to sweep through the town built on Valdez Glacier's flood plain, he chose "Ludington's Landing" on the southeast side of Port Valdez because it had a good water supply for sanitation purposes, an ample wood supply, a place for a rifle range and was close to a good anchorage while

After experiencing the severe floods of 1899, Abercrombie concluded that the townsite of Valdez was "utterly valueless of any purpose whatsoever" and selected a military reservation across the bay from Valdez. Ft. Liscum was built in 1900. Today, the Alyeska Marine Ter-minal occupies this site. Photograph No. B62.1.194 reprinted courtesy of The Anchorage Museum of His-tory and Art.

far enough away from Valdez "to be beyond the influences of the whisky element to be found in frontier towns (p. 33)." It was also the area Gillette felt most suitable for a railway terminus. In May of 1900, building began on "Ft. Liscum." (named for Col. E. H. Liscum who was killed in the Boxer Rebellion in China). By the fall, construction was completed on twenty-one buildings including an officer's quarters, stable, bakery, hospital, several storehouses and barracks housing 109 soldiers.

Fort Liscum contributed to the town's economy, political and social life. Lt. Rafferty, as already noted, was a member of the first township committee and a number of soldiers took out townsite lots. Ft. Liscum's Tillikum Club, the officer's club, hosted many political and social gatherings; and Abercrombie and his officers were frequent guests at parties in Valdez.

The All-Alaska Telegraph Line

Ft. Liscum served as the army's headquarters for constructing and maintaining the All-American road and the Washington-Alaska Military Cable and Telegraph System (WAMCATS). In 1899, the War Department initiated plans for a telegraph line for military communications and commercial business within Alaska. In 1900 Congress approved $450,000 for its construction. It was only natural that the telegraph line follow the military road from Valdez to Ft. Egbert near Eagle City. Even before its completion, the telegraph aided the miners and radically changed communications along the trail by providing same-day coverage of the weather, trail conditions and locations of parties on the trail as illustrated by this first newspaper report received by telegraph:

> Tonsina. Apr. 12. The last of the outfits bound for the Chittyna are now well on their way down the river. Amy's outfit and others are somewhere near Copper River but as no one has yet come back up the river it is not known how they are getting on.
>
> Copper Center. Apr. 12. The ice on the rivers in in excellent shape for sledding. Most of the outfits have passed this place, some of them are reported to be as far as the Talsona river. Nights have been cold thermometer registeringas much as 20 below zero. Capt. Barnell and Lieut. Mitchell are guests at "Hotel Holman."

Telegraph Station No. 4 from Valdez on the All-Alaskan Telegraph line or WAMCATS. Photograph by the Miles brothers.

From the Miles Collection, courtesy of the Valdez Museum and Historical Archives.

Station Three, Apr.12. The trail at this place is in excellent condition. The canyons between here and Stewart Creek are reported to be in very good condition. Thermometer registered 10 below zero this morning. Jones brothers and a few other outfits are campd at the head of the canyon. (By Telegraph: *The Valdez News* 2:5. 4/12/02, p. 3).

The line was completed on August 24, 1902 when Lt. William "Billy" Mitchell and Lt. George Burnell met at Tanana Crossing (now Tanacross). With the completion of the line, military personnel, mining developers, railroad magnates and others could communicate around the world within a few hours. Valdez became the transportation and communication center for southcentral Alaska.

Gold, Valdez, and Memories

Although many would have been skeptical during the fall and winter of 1898, by the end of the summer of 1899, the future of Valdez as a transportation corridor to interior Alaska was secure. An All-American route had been found and part of the Trans-Alaska Military Trail along that route constructed. By 1901, Valdez was a thriving town. Townsite lots claimed for $2 in 1898 now sold for $100 or more. The discovery in 1902 of rich placer deposits near what is now Fairbanks would add further impetus to improving this route. When the gold mines near Fairbanks began production some time later, the major route to the golden interior of Alaska led through Valdez. Between 1905 and 1907, miners and prospectors, gamblers and con-men, dancing girls and prostitutes, saloon keepers and other merchants, judges, attorneys, and politicians all arrived in Valdez. Law and order became a major problem. Valdez had finally achieved the status of the classic gold rush town.

However, this gold rush and this Valdez were a far cry from those known to Bourke and Guiteau, to Austin, Remington, and Townsend, to Moore and Margeson and all the others who braved the Valdez Glacier Trail of '98. The riches that this rush brought into the lives of these men and women cannot be measured by the gold standard. Margeson reflects on this on leaving the Sound during the fall of 1898.

> At early dawn we were on-our way out of Prince William Sound, and taking the outside course, the snow-capped peaks of Alaska soon disappeared in the distance.
>
> We will not try to describe the mingled emotions with which we watched them disappear. These mountains, and rivers, and glaciers, and lakes, the hardships and perils, the pleasures and pains, the hopes, anxieties, ambitions, and disappointments which had been crowded so thickly into the past months, were all among the things that were past, but not all among the things to be left behind.
>
> Many, aye, the most of them, were to come back with us to the States as very lively memories, to go with us all the future years, to sometimes be lived over in reminiscences and story, or in dreams of the night (Margeson, p. 293).

Roll Call

Contemporaries estimated that between three and four thousand came to Valdez in 1898-99. The following list represents about a third of these. Their names appear in mining district records, military and other government reports, diaries, township minutes and property records, letters, news paper articles, the Crary and Bourke scrapbooks, and in journals and books published by the participants. If you have information on a person who participated in the Valdez Gold Rush episode, we encourage you to contact the Valdez Museum and Historical Archives.

Men of '98-99

—, Gus (German crewmember on Hera)
Abercrombie, Capt. W. R.
Abercrombie, J. J.
Ables, J. W.
Ackley, —. [Spirit Lake, Iowa]
Adams, A. W.
Adler, M. L.
Ahren Brothers
Alleman, —
Allen, Dr. —
Allen, J. C. [Fall River, MA]
Allis, H. G. [Little Rock, AR]
Alstrum, James [Highwood, CT]
Ames, Eugene G.
Ammann, Adolph [Catskill, NY]
Amy, W. S.
Anacker, W. H. (Also Anaker)
Anackin, Wm. H.
Anderson, — [Boston, MA; Sweden]
Anderson, Charles
Anderson, Ed
Anderson, Gus
Anderson, J.W.
Anderson, Ole
Anderson, P.

Archer, Private —.
Arden, B.
Arnold, H.
Arnold, H.W.
Arnold, James W.
Arnold, W.
Ash, Ed
Ayers, —.

Baasen, R.V. [MN]
Babcock, First Lt. Walter C.
Bacon, —.
Baird, Jim [Ellensburg, WA]
Baird, R.S.
Baker, C.G.
Baker, James S.
Baker, W.S.
Baldins, E.M.
Bargeson, John
Barker, Charles [Sacramento, CA]
Barker, W.
Barnes, F.C.
Barnhart, M.S.
Barnum, W.P. [Beaver Falls]
Barratt, Doc.
Barrett, Dr.
Barrett, June J.
Barrey, —
Barrington, —.
Bartlett, Frank G.
Bartlett, H.D.
Bartlett, S.J.

Bates, Ben
Beach, Irving
Beatson, A.K. [Oakland, CA]
Beatty, H.
Beatty, H.T.
Beatty, J.T.
Beatty, L. [San Jose, CA]
Beatty, Mike A. (Also Beaty, M.A.) [MN]
Beatty, S.T.
Beck, Jack
Becker, Bill [Freeport, IL]
Behren (?), Chas. F.
Behrens, Henry [Brooklyn, NY]
Bell, —.
Bell, James H.
Bell, John
Bell, L.
Bempton, Aug.
Bench, Private —
Benedict, F.L.
Benedict, Neal D. [Seville, FL]
Benjaman, —.
Bennett, C.F.
Bennett, C.L.
Benson, —.
Benson, Peter
Benson, P.A.
Bent, H. [MA]
Berg, Charles
Bergmann, J.S.

Beringer, Fritz
Bettles, Jas.
Betts, C.L. [Spokane, WA]
Beyer (s), William Carl Luckner [Germany]
Birch, Stephen
Bjornstad, Martin B.
Black, Arch
Black, J.C.
Black, Jim
Blackman, —.
Bland, Robert
Blanding, W.D.
Blix, Ringwald [Norway]
Blum, L. (or S.)
Blumauer, Louis
Blumauer, Phil (Also Blaumer, Phil)
Bogen, Bill
Bohnsen, J.M.
Booker, Ben
Borden, —.
Borgeson, John
Bornd, John
Borud, J.
Bourke, Joseph (Burk, Bourk) [Brooklyn, NY]
Bowen, Tom [Texas]
Bradburg, J.H.
Bradbury, J.W.
Brandenburg, J.T.
Brandenburgh, E.T.
Brandenbury, G.W.
Bradford, J.H.
Brandon, —.
Brehme, Chas. F.
Brehme, Joe
Breine, Chas.
Brook, K. Edward
Brook, William [Stamford, CT]
Brookfield, Second Lt. R. M.
Broughton, Rob
Brown, Charley
Brown, George
Brown, L. ("Farmer Brown")
Bruce, — [Pittsburgh, PA]
Bruce, Walter C. [Tacoma, WA]
Buckley, J.H. [St. Peter, MN]

Bulger, James
Bulger, John K.
Bulls, Charles H.
Burns, W.T.
Busby, W.J.
Bush, Lewis
Buskirk, W.H.
Buskirk, J.P.
Butler, Harry [Carinda, Iowa]
Butler, H.N.
Butler, Jesse [Carinda, Iowa
Butts, Charles [Stamford, CT]
Byjorum, —

Cahill, — (Also Kahill)
Call, N [Worthington, MN]
Cameron, Fred J
Camicia, Dr. L.S.
Campbell, F.A.
Campbell, P.
Carlin, B. ("Uncle Jim") [MA]
Carlon, J.R.
Carlson, August
Carlson, Olaf (Also Carlsen, Olaf)
Carlson, Tening
Carroll, Jack
Carvey, Bert
Carvey, L.H.
Caryl, W.T.
Cashin, Chas.
Cashin, D.A.
Cashin, Thomas D.
Cashman, Edward Peter
Chamberlain, C.W. [Tacoma, WA]
Chesterman, Hank M.
Chisholm, Angus (also August Shesholm)
Chisholm, W.J.
Choudichild, John
Christiansen, Chris (also Christianson)
Christopher, Dick [Brooklyn, NY]
Clancy, Tom
Clas, Tom
Cleave, Corporal John W.
Cloudman, Peter
Cockerille, Dr. — [Washing-

ton, D.C.]
Cohn, — [New York]
Coke, —
Colby, — [Sweden]
Cole, Bob (also Robert H. Coles)
Coles, Howard (also Cole)
Colis, R.H.
Collins, Charley
Collis, Bill [Tacoma, WA]
Condon, —
Cone, S.J. [Litchfield, CT]
Conger, Horace S. [Mora and Kasota, MN]
Cooley, —
Cooper, —
Cooper, E.J.
Corcoran, John [MN]
Corless, Fred
Corlis, Fred (also Ed Corlis)
Costa, Jack ("Happy Jack")
Cramer, A.F.
Crames, A.T.
Craquet, Capt. —
Crary, Chas. N.
Crary, Will. H.
Crawford, A.K.
Crawford, Archibald (also Archie)
Creek, D.B.
Creighton, Col. —
Crockett, Capt. —
Culver, H.M.
Curran, John
Cushman, Dan

Dable, John
Dahl, Pete
Dankert, Charles
Date, Fred J. (also Fred E. Date) [Elkhart, IN]
Davidson, J.C.
Davies, Daniel T.
Davis, —
Davis, D.T.
Davis, George F.
Davison, George
de Fontville, Ph.
Delander, J.
Dempsey, Melvin

Denby, Wythe
Denler, Geo.
Denonin, Antonin
Dermham (?), D.
Dermham, H.
Dexter, John
Dicker, L.A.
Dickerson, Sam K.
Dickey, — ("One-armed Dickey")
Dickey, George
Dietrick, —
Dinham, F.H. (Durham?)
Dinsmore, Dr. G. B. [Peoria, IL]
Diston, John
Dittman, Paul, M.D. (also Ditman, Dr. —)
Dittman, R.B.
Djarf, Gus
Dolan, —
Dolan, Tom
Dolloff, Frank
Donaldson, —
Donaldson, G.W.
Dooley, Geo.
Dorsey, —
Dowling, Albert C. [Dublin, Ireland]
Downey, John (also Downing, J.C.)
Downing, J.C. (also Downey, John)
Downing, J.R.
Drace, —
Drace, Kit
Drake Brothers
Drase Brothers
Drew, —
Du Bourdien, James (Du Bourdim?)
Duff, Clark
Dufresne, J.A.
Dum, George
Dunham, D.F.
Dunham, F.H.
Dunham, Howard
Dunn, Geo.
Dunn, Harry
Durham, F.H. (Dinham?)

Dyette, —

Eagins, J.R.
Eaton, John R.
Eblerkamp (see Ellerkamp)
Eckland, John [Haywood, WI]
Edler, Axel B.
Edward, Dr, —
Edwards, — ("Blackleg Edwards")
Edwards, Thos. E.
Egerter, W.H.
Egram, John
Ehrhardt, Adolph [New York]
Eighrny, — (also Eighmy) [San Francisco, CA]
Eighmykn— (also Eighrny)
Elder, Axel B.
Ellerkamp, Rudolph (also Eblerkamp) [Louisville, KY]
Elliman, Alfred (also Alleman) [New York, NY]
Emerson, Thomas
England, Axel
Erickson, —
Erickson, Carl O.
Erickson, Ed. A. (Errickson, E.A.)
Erickson, Julius (Errickson, Julius)
Erickson, L.B.
Ericson, John E.
Errickson, E.A. (also Erickson, Ed A.)
Errickson, Julius (Also Erickson, Julius)
Evyan, — [Norway]

Faber, —
Fain, Harry
Farnsworth, Capt. —
Farrell, John
Fawcett, A.V.
Feldman, Jo
Fenske, F.E.
Fenske, H.E. (probably Fenske F.E.)
Ferguson, J. B.
Ferguson, Robert P.

Feyne, Harry
Fields, —
Finch, Jim W.
Finical, William
Fischlein, — (also Fishline) [Brooklyn, NY]
Fish, P.A.
Fisher, —
Fisher, Samuel [Chicago, IL]
Fishline, — (also Fischlein)
Fitch, H.H. [Great Falls, MO]
Fleming, —
Fleming, G.W.
Flemmings, —
Floam, Albert C.
Flynn, —
Fogg, George
Fohlin, John
Folan, Gunan
Folk, W.H.
Forbs, J.E.
Forshner, Charley [Rhode Island]
Foster, —
Foster, Geo. A.
Fournier, Joseph (Chicago, IL]
Fox, J.C.
Frain, O.H.
Fram, Party (6) [Scandinavia]
Frase, Frank
French, W.W.
Frickel, Valentine [Orange, NJ]
Fritz, Colonel —
Furgusen, Robert
Fury, John H. (Also Furey, John) [Milwaukee, WI]

Gadt, Charles
Gage, L.F.
Galusha, B.T.
Gannett, Fred C.
Gardner, William J.
Gardiner, S. (also Gardner, S.)
Gardiner, John (also Gardner, John)
Garner, A.B.
Garnes, A.B. [Roslyn, Washington]
Garrett, James W. (also Gar-

ret, James W.)
Gasteldi, Bernard [Norwalk, CT]
Gatd, Charles
Gates, E.A.
Gates, Herbert B.
Gates, H.L.
Gelineau, Dick (also Gelene-au, Dick)
Gelineau, J.K.
Geran, G.F.
Getz, G.
Gillespie & Co.
Gillett, William L. (also Gillet, William L.)
Gillette, —
Gillette, Edward
Gittner, Fred [Cannons, CT]
Gladhaugh, M.O. (also Glad-hough, M.O.)
Gleason, A.D.
Glendenning, —
Glenn, Edwin F.
Gliason, Arthur D.
Golden, C.H.
Goodell, H.
Goodell, M.E.
Goodman, W.E., Jr.
Gordon, J.S.
Gorich, J.A.
Goss, P.B.
Gott, C.
Grady, J.E.
Grasser, John
Gray, —
Gray, Smith W.
Greene, J.D.
Greenig, Dan T.
Grey, S.J. (?)
Griffith, —
Grinnell, J.S.
Griswold, Park (also Griswald, Park)
Grogg, Sylvester
Grogg, Wm.
Groo, Samuel J. [Salt Lake City, UT]
Grumaer, —
Grund, August [MN]
Guiteau, Luther ("Lute")

Gunstat, M.
Gutherie, G.T.
Guyer, Frank [MN]

Haber, George
Haberstroch, J.E. [Chicago, IL]
Hackett, J.E.
Hafter, Andrew
Haines, B.F.S. [MA]
Hall, A.G.
Hall, James [Stamford, CT]
Hallett, Private —
Halt, C.C
Ham, Sam
Hamlen, J.E. (also Hamlin, J.E.)
Hanalin, J.
Hannam, J.L.
Hannam, J.H.
Hansen, —
Hansen, E.
Hansen, George
Hansen, Harry
Hanson, —
Hanson, —
Hanson, Garrett
Hanson, Johnny
Hanson, Otto
Hanz, —
Hardwick, E.N.
Hardy, F. (?) W.
Harrington, F.B.
Harrington, Peter
Harris, Frank
Hartley, Clark ("Shortie")
Haskell, Frank H.
Hatch, G.A.
Hauschild, —
Hawkins, R.S.
Hayden, Jack
Hazelet, George Cheever (also Hazlett) [Nebraska]
Hedman, J.W. [Chicago, IL]
Hegland, Nelse [Appleton, WI]
Heiden, Robert (also Hyden, Robert)
Heidorn, W.B. (also Heidoors, W.B.)

Hemmings, Geo. W.
Hemple, S.A. ('Oklahoma Bill')
Hemple, W.S.
Hemrick, Louis
Hendricks, W.E.
Hendrie, A.M.
Henkel, John
Hennesy, W.M.
Henning, Wilbur [Braddock]
Henninger, Fred L.
Henry, F.A.
Henry, F.H. (poss. = Henry, F.A.)
Hermann, John R.
Hernaruk, (?) Alvin
Hernaruk, (?) Louis
Heron, R.T.
Herriott, Geo.
Hertel, L.
Hertzberg, H.
Hessian, Mike J. [MN]
Hessian, Neil [MN]
Hettler, Bart
Hickman, —
Hildreth, H.H. [Massachu-setts]
Hill, —
Hill, Charles P. [Colorado]
Hill, H.H.
Hinky, Al
Hineky, —
Hoagland, J.M.
Hoffman, —
Hoffman, A.F. (Frank)
Hofler, Andrew [Dorchestr, WI]
Hogan, T.D.
Hogg, George (also Hogue) [Newsburgh, N.Y.]
Holland, —
Hollingsworth, C.T.
Hollingsworth, C.F. (prob. = C.T.)
Holman, A. [Norway]
Holmes, Peter W. (also Holm-es, Pert W.)
Holt, C.E.
Holton, O.B.
Homer, A.

Homer, Frederick
Hopkins, Wm. F.
Horn, William
Hosford, O.J.
Hoy, L.D. [Seattle, WA]
Hoyt, Frank W. (also Hoit, F.)
   [Norwalk, CT]
Hubbard, Chas. H.
Hubel, L.
Hudson Boys
Hudson, Dr. —
Humphrey, Capt. Omar J.
Humphrey, H.N.
Hunt, H.E.
Hunt, P.S.
Hunt, Thomas H.
Hunt, W.E.
Hunter, —
Hurd, Bert (Hurd, O.B.)
Hutchinson, C.S.
Hutchson, Shuman
Hyden, Corporal — (also
   Heiden, Corp. Robert)

Iams, J.
Ikes, James
Iliff, H.G.
Ingram, J.
Irwin, C.H.
Isert, Rob (Also Izatt, Robert)
   [Milwaukee, WI]
Izatt, Robert [Milwaukee, WI]

Jack, —
Jackson, Peter
Jacobsen, —
Jacobsen, J.
Jacobsen, Nils John (also Ja-
   cobson, Nils John)
Jacobson, —
Jaffey, —
Jaknert, —
Janonski, John
Jarks, T.J. (Also Jerry Jarks,
   Tjarks, T.J.)
Jaworsky, J.A.
Jaworsky, Wesley [Hornells-
   ville, NY]
Jefferson, Jesse Delos
Jeffries, Lt. — (See Jefferies)

Jenning, —
Jennings, H.F.
Jensen, George
Jenson, A.S.
Jepson, Nelse J. [Sweden]
Jilsan, A.T. (possibly, Jilson,
   A.T.)
Jilson, A.
Johnson, —
Johnson, Charlie
Johnson, D.
Johnson, Edward
Johnson, G.H.
Johnson, Gus
Johnson, John A.
Johnson, Ole
Johnson, Oscar
Johnson, Swan
Johnson, T.M.
Johnston, Nelse
Jones, Alfred W.
Joy, Charles R.
Joy, Frank S.

Kaher, Joseph F.
Kahill, —
Kain, Dan S.
Kaiser, Herman (also Kiser,
   Herman)
Kane, —
Kanitz, Otto
Kantens, (?) George (Kautens
   or Kauteus)
Kartright, Dr. S.E. (also Kor-
   tright, Dr. S.E.)
Kaurin, Odin
Kehoe, Henry (also Kehor,
   Henry)
Keller, J.W.
Kelley, Charles [Providence,
   RI]
Kelmick, Henry
Kem, George (or Kene ?)
Kendrick, H.B.
Kendricks, —
Kernan, Frank
Kertchum, Frank (Also
   Kertchem, F.) [Portland,
   OR]
Kietzel, Herman

Kimball, Fred
King, Harry E. F. [Stamford,
   CT]
Kingsley, Ed [Freeport, IL]
Kirkham, R.V. [Montana]
Kirkpatrick, — [Ellensburg,
   WA]
Kiser, Herman (also Kaiser,
   Herman)
Kitcher, Henry [Bridgeport,
   CT]
Kloeber, Chas. E.
Knott, —
Koehler, Corporal Robert
Koehler, Karl [Juneau, AK]
Koppus, Chas. (also Kopus,
   Chas.)
Kortright, Dr. S.E. (also Kar-
   tright) [Hoboken, NJ]
Kowasky, A.
Kraemer, Charles H.
Kraft, Dr. F.A.
Kraft, Fred
Kraft, Rudolph
Krau, George [Elkhart, IN]
Krohn, Henry
Krumholz, Fred
Kulman, C.H.
Kulper, Hein
Kyberg, Gus
Kyberg, Ralph

La Sage, Chas.
Lane, H.
Langkabel, Otto
Langworthy, J.O.
Larson, E.S.
Larson, Oscar
Larson, R.
Larson, Ray
Lashaway, Hy
Latterner, George (probably
   also Leturner, Laterner,
   La Turner)
Lavalle, Judge —
Lavarar, Frank
Lavell, H.S. (also Laull, H.S.)
Lavigne, Frank
Lavioss, J.
Lawrence, F.C. [Ashcroft,

B.C.]
Lawrentz, W.H. [Litchfield, CT]
Lawson, Andrew
Lawson, Joseph [Norwalk, CT]
Laymer, Julius (Also Leymer, Julius)
Leach, — [Lyons, NY]
Leap, Fred O.
Learndas, B.F.
Leavell, Dorsey (also Levell, Dorcey, Leavelle, Dorsey)
Lee, Oliver [Subley, IA]
Lee, W.A. [Lowell, MA]
Lehn, Richard
Leliners, Carl F.
Lennet, H.
Lennore (?), F. Ed.
Leonard, W. L.
Lesig, Cael
Lesig, H.R.
Leveroos, B.G. (see Louveros) [Minneapolis, MN]
Levy, A. L.
Lewis, A.
Lewis, Capt. Dick
Lewis, Dr. —
Lewis, William
Leyshon, John Graham
Liljegren, Oscar
Limoseth, P.
Lind, Axel W.
Lind, C.A.
Lind, C. H.
Lind, H.V.
Little, Joseph
Lockridge, —
Lofgreen, —
Logan, Dr. —
Lomnes, E.L.
Looken, E.T.
Lorraine, J.O.
Love, G.W.
Love, Homer L.
Lowe, Lieut. P. G.
Ludwig, S. J.
Ludwig, W. G.
Lusk, Frank

Lynch, Samuel S.
Lyne, Wm.

Mackin, John
Mahler, George
Mahlo, Emil
Maine, — [Spirit Lake, Iowa]
Makiesport Party (Also Mckeesport)
Malley, Bill
Malley, Josh
Malmgren, Victor T. [Chicago, IL]
Malone, Lon J.
Mam, William
Mambock, Julius B.
Mamon, Charley
Mamon, Thos.
Manker, James, A. [Hamilton, MA?]
Mann, Wm.
Marforden, —
Margeson, Charles A. [Hornellsville, NY]
Marshal, J.C.
Marshall, J. R. [Mexico, MO]
Martin, James
Martin, W.B.
Mass, Jacob
Matthew, Dr. — (Also Matthews, Dr.)
Mattson, Matt (Also Matson, Matt)
Maxwell, J.E.
Mayhew, —
McCabe, Duncan (Also MacCabe, Duncan)
McCall, Charles
McCall, W.C.
McCarthy, James (Also, McCarty, James)
McClary, —
McClellan, Dave
McClellan, R.F.
McConnell, W.B.
McCormick, Frank [San Francisco, CA]
McCoy, E.L.
McCoy, C.E.
McCracken, —

McCullough, Dan
McDonald, A.M.
McDonald, Alex S. (Also, McDonald, Al)
McDonald, John
McDowell, A.C.
McDowell, A.S.
McDowell, Bert
McFall, N.A. (Also McFaul, N.A.)
MacFarland, —
McFarland, Thos. E.
McGee, Harry
McGee, Jack
McGee, T.D.
McGee, W.C.
McGee, W.D.
McGee, W.J.
McGrew, —
McHie, James
McIntyre, —
McIntyre, John
McKeag, George
McLeman, A.S.
McMillan, —
McMillan, E.A.
McMullen, George B.
McNair, Walton D. (Also McNear) [Sitka, AK]
McNair, William H. (Also McNear) [Sitka, AK]
McNamara, Jack
McNear, Arthur H. [West Virginia]
McPherson, Capt. James
McSheehy, J. B.
Meals, Andrew Jackson
Means, James
Means, Sam. B.
Meislang, —
Meiss, Ed
Melby, E.G.
Melby, Ed
Mellark, Henry (also Mellerk, Henry)
Melley, E.G.
Mendins, Henry
Menke, Emil
Merchant, Norman
Merrifield, George

Messer, C.J.
Meyers, Ed
Michaelson, Dick
Michaelson, Robert
Mieslang, Thos. A.
Mightener, Louis
Millard, Benjamin Franklin
Millard, Ray
Miller, — ("Old Man Miller,")
Miller, —
Miller, Henry
Miller, Jack
Miller, John
Miller, Maximillan [New Yor]
Miller, William [Carpenteria, CA]
Miller, Wm. T.
Milligan, S.
Mills, S.J.
Mondard, —.
Montgomery, Jim
Moon, Harry
Moore, Chrs.
Moore, Clark [CA]
Moore, H.L.
Moore, Otis
Moore, S. C.
Morgan, Charles
Morgan, David
Morgan, H.L.
Morrison, Joseph
Morse, Gus
Morse, S.A.
Morton, John
Mourer (?), George, C.
Moyes, Capt. Emanuel [Stamford, CT]
Moylan, Patrick
Murfin, Le Roy (Also Murfine, Murphy) [Sleepy Eye, MN]
Murphy, D.T. [Stamford, CT]
Murphy, J.J. [MA]
Myers, F.M.
Myra, Ed
Mytinger, L.J.
Nelson, —
Nelson, Andrew
Nelson, C.
Nelson, C. G.

Nelson, John
Nelson, M.
Nemo, —
Newberg, —
Newcomb, J.
Newcomb, Jim P.
Nichols, J.
Nicholson, James M.
Nierman, Dr. H.G.
Nitschie, V.H.
Nokes, John ("Dad") [Arkansas]
Nolan, — [Jefferson City, MO]
Nolan, W.J.
Noran, Chas. (Also Noren)
Norn, Fred
Nucum, James
Nuskolls, Richard (also Nuckolls, Nuskalls)
Nutter Brothers
O'Brien, M.D.
O'Brien, Thos.
O'Connel, D.
O'Connell, Daniel [Glenbrook, CT]
O'Connell, Noah
O'Conner, Neal
O'Hara, Tom (Also, O'Hare, Tom)
Oberfeld, Adolph [Booneville, MO]
Ohlemutz, H.
Ohlhausen, A.
Ohlhausen, Alex C. (also Ahlhausen)
Ohlsson, Nils E. ("North East") (also Ohlson)
Oleson, Arthur P. (also Olson)
Oleson, Edward
Oleson, Ole [MN]
Olsen, Fred
Olsen, Otto R.
Olsen, Tom (Also Olson, Tom)
Olson, Arthur P. (also Oleson)
Olson, John L.
Olson, Oscar
Olson, Svend [Northwood, ND]

Onstad, O.H.
Onstead, Wm.
Opdahl, T.H. (Also Updol, Richard; Ophedal) [Marshal, MN]
Opie, Harry
Orr, W.H.
Osborn, Fred [Lynn, MA]
Osgood, D.F.
Osgood, W.F.
Osland, —
Otaway, Dr. — (also Ottawa) [Rochester, N.Y.]
Otterness, J. A.
Owens, F.R.
Oxelgren, J.A.
Page, C.H. [Buffalo, NY]
Palak, —
Palmer, —
Palmer, —
Parsons, Job
Parsons, W.H.
Patterson, Jas. H.
Patton, A.O.
Patton, Geo. W.
Paulsen, —.
Paulson, Paul
Payne, J.M.
Payne, K.C.
Pearson, Elmer
Pearson, Dr. H. Brian (also Bryan and Pierson)
Pearson, Orie S. (also Pierson, Ore)
Peary, T.W.
Peck, —
Peckham, Charles H.
Pedersen, Cornelius ("Bald-headed Chris," also Pederson, Peterson)
Peets, H.G. [California]
Perry, Frank W.
Peters, David T. [Chillicothe, MO]
Peterson, B.F.
Peterson, B.O.
Peterson, C.
Peterson, Oscar H. (Also Petterson, O.H.) [Chicago, IL]

Petrie, Jack
Pfennig, Ernst
Phelps, Albert
Phillips, W.S.
Pierson, Dr. — [Chicago, IL]
Pierson, Ore (also Pearson, Ore)
Pinkham, Jack
Pittock, K.
Pitts, C.L.
Place, J.V.
Polinsky, John
Pollard, Robt. H. [Council Grove, KA]
Polley, Wm.
Pollock, Mark [Omaha, NE]
Polowitch, —
Pope, E.A.
Pottes, —
Potts, John [Westport, CT]
Powell, Addison M.
Preasley, J.B.
Pressley, Dr. —
Preston, First Lieut. Guy H.
Priceler, Charles [Denmark, NY]
Proden, —
Putz, Joseph
Quick, Dr. —

Rafferty, Lieut. J.J. (Also Raffety, Lt.)
Ramsdell, M.
Randall, F.H.
Ranous, L.P.
Rape, E.
Rayl, B.W. [Beaver Falls. PA]
Reed, Robt.
Reigel, J.H. [MN]
Remington, Charles H. ("Copper River Joe") [MN]
Remmington, Grant H. [MN]
Reonolds, —
Reynolds, C. C. [Detroit, MI]
Reynolds, John [Stamford, CT]
Ribbstein, Billy (Also Ripstein, W.E.)
Rice, B.H.

Rice, Ira
Rice, John F.
Richards, Hiram
Richards, Rev. O. A.
Richards, Judge J.B.
Riggins, C.H.
Riggins, J.M.
Ripstein, William (also Billy Ribbstein)
Robe, Harvey A.
Robert, Willet
Roberts, —
Roberts, Frank [Salt Lake, UT]
Roberts, Jas. P.
Roberts, J. R.
Roberts, Wm. H.
Robertson, W.
Robeson, W. Neil (possibly Robison)
Roff, — [Salt Lake, IO]
Rogan, Jas.
Rogers, T. O. [Danbury, CT]
Rohn, Oscar
Romdell, Frank
Rope, —
Rorer, George [New York]
Rosenthal, F.W. ("Tenas Rosy,") [Missouri]
Rosenthal, ?. S. [Laport, IN]
Roser, Ed
Ross, Dan
Rothkranz, Louis (Also Rotheranz, Louie, Big Lewy Rothcranz)
Rowley, Luke
Royer, L.A. Tony [New York]
Rua, Charles T.
Rude, H. P.
Ruggs, —
Russell, Carl
Rutherford, Roy

Saches, Carl (Also Sachs)
Sand, Frank
Sands, John
Savage, Walter
Sawyer, C.A.
Scheime, Alfred (Also Schieme, A.)

Schelly, —
Schilling, Mike
Schimmer, J.W.
Schlagel, — (Also Schlogel, Schlagle, Schloegle)
Schloegel, F.W.
Schlosser, Charles (Also Sclosser)
Schnedeger, Wesley
Schnettgocke, Aug.
Schnyder, J. P.
Schrader, Frank C.
Schroder, W. A.
Schultz, Adolph
Scott, Andrew [MA]
Scott, Fred
Scott, John O. [MA]
Scott, Peter
Sebastian, —
Sender, Isaac (also Sender, Ike)
Sharp, — [CO]
Sharp, — [Chicago, IL]
Sharp, George [Rosebud, MO]
Sharp, T. [Braddock, ?]
Shaw, H.C.
Shaw, Percy [Springfield, MA]
Shaw, Sidney [Springfield, MA]
Sheehy, Jas. B.
Sheiner, J.H.
Shelly, P.W.
Shelton, L.T.
Shepard, Jack (Also, Shepard, W.J.)
Shiennie, Alfred (Also Shiennie, Alfred)
Skinner, Bert
Shipp, Frank
Shultz, Robert
Simensted, Chas. (Also, Simonstead, Simenstod)
Simonson, Anton [Chicago]
Simpson, Jas. [Bridgeport, CT]
Simpson, W.H.
Singletary, Alex
Singletary, R. L.
Sittell, Jacob [Portland, OR]

Skeddy, William
Skully, — (Also Scully)
Slattery, Pat
Smith, — ("Middleton Island
    Smith")
Smith, A.D. [WA]
Smith, Charles B. [Lanesboro,
    PA]
Smith, Dr. —
Smith, F. M.
Smith, Flavel
Smith, Fred
Smith, Hardin
Smith, Heber [New York, NY]
Smith, Harry T. [St. Paul,MN]
Smith, I.N.
Smith, J.A.
Smith, J.D.
Smith, J.W.
Smith, Jack ("Arizona Jack")
Smith, John H.
Smith, M.N.
Smith, Mike
Snap, Wendell [CA]
Snow, Philo
Snyder, Joe
Snyder, John Grant
Speddy, Wm.
Spencer, —
Spencer, Wm.
Spongberg, Charles A. (Also
    Sporngberg, Sponberg)
Spotts, —
Sprague, D.W.
St. Clair, Charlie
Stahly, I.
Stark, Jim [Australia]
Starny, Thomas
Stayer, Wm.
Stead, Jake
Stead, Lindsey [Sound Beach,
    CT]
Stead, Phillip [Sound Beach,
    CT]
Stebbens, John
Steel, James [Chicago, IL]
Steel, Robert
Stehn, John [Benicia, CA]
Steinmutz, P.
Sternberg, J.

Stevens, Barney
Stevens, Dal.
Stevens, S.C.
Steward, Frank
Stewart, H. M. [Rochester,
    NY]
Stewart, Jack [Rochester, NY]
Stewart, ?, M.
Stiles, Wm.
Stodart, J. [Osceola Mills, PA]
Storey, Walter
Stotter, John (Also Stoter)
Strode, Jack
Strunk, Antoine ("Dad")
Stryers, Wm.
Studt, Private —
Suhen (?), Jacob
Summers, W.J.
Sumner, —
Sumner, Clement Moore
Surrt, W.W.
Suter, Jacob
Sutter, —
Swager, Private —
Swan, Adam
Swanson, Charles [Peoria, IL]
Swanson, Charles J.
Swanson, Joe
Sweeny, John F.
Sweet, —
Sweet, H.H. [Hornellsville,
    NY]
Swensen, J.P.
Swenson, L.
Syring, Herman
Tak, James (Also Tek, James)
Tanner, Millard Filimore
    ("Doc") [Lexington, KY]
Taylor, Bayard
Taylor, Rev. Sam W. ("Whis-
    key Taylor")
Teeters, Capt. —
Thiery, Joseph C. [Chicago,
    IL]
Thompson, Harry [New York,
    NY]
Thompson, William
Thormudson, T.
Thornell, J.
Thorstensen, Louis (Also

Torsenfen, Louey)
Thuland, C.M.
Tibbitt, A.W.
Tiner, Bill
Tisdale, John N.
Tjarks, T.J. (Also Jery Tjarks,
    Jerry Jarks)
Tjosvig, Christopher (also
    Tjosvig, Christian)
Tohlon, Gunnan
Tolley, Private — (Also Tully)
Topping, B.H.
Torgeson, Hans
Tottensen, M.L.
Townsend, Dr. Leroy
Tracy, E. J.
Traut, Paul [Peoria, IL]
Traveland, Christopher [Eure-
    ka, CA]
Treet, Geo. C.
Treloar, William Elmer
    [Carpinteria, CA]
Trevdt, Phil
Trew, Dr. Neil C.
Tripp, —
Trook, J.
Tully, Private — (Also Tolley)

Updahl, T.H. (Also Opdahl,
    T.H.)
Updike, L.
Updyke, Jr.
Uppercue, —
Uppercue, L.
van Antwerp, B.
Van Court, Elias
Van Sant, —
Varley, Pat [Minnesota]
Veazey, Simon
Vogel, Lutz
Voight, Richard (Also Voigt),
    [Norwalk, CT]
von Ciple, Count —
Von Gunther, Dr. A.

Wadman, —
Wadman, A.J. [Glasgow, MT]
Wagey, A.M. ("South
    Dakota")
Wagner, William [New York,

NY]
Walker, E. Fox
Walker, Ernest ("Whiskers") [England]
Walker, Frank B.
Wallace, —
Wallace, N.
Walsh, Ed.
Wann, —
Ward, Joseph B.
Warner, Clarence L.
Washburn, —[Fitchburg, MA]
Waterman, —
Watson, —
Watson, A.W.
Wayland, —
Webster, — [CO]
Webster, Dr. —
Webster, H.
Webster, John (also Jack)
Weckert, Joseph. V. Jr.
Wehrley, A. [MN] (Also Whirley, Andrew)
Wehrley, D. [MN] (Also Wherley, D.; Whirley, Druey)
Weiler, John
Weinnart, Doc.
Weir, Henry A.
Weiss, G.A.
Wells, Frank
West, Capt. I.N.
Westervelt, —
Whaler, J.H.
Whan, Nick
Wheat, Ernest B. [WA]
Wheeler, N.D.
Wherley, D. [MN] (Also Whirley, Druey)
Whipp, John
Whipple, John F.
Whirley, Druey (Also Wehrley) [MN]
Whirley, Andrew (Also Wehrley) [Belle Plaine, MN]
Whitcomb, Enock W.
White, Al
White, E.A. (Check)
White, M. (Check)
Wiggs, H.K.

Wilcox, —
Wilkins, Charles
Wilkinson, —
Williams, Issac
Williams, Jay
Williams, John
Williams, Williams [Stamford, CT]
Wilson, Dan
Wilson, H.G.
Wilson, John A. [WA]
Wilson, M.
Wilson, W.A. [Brooklyn, NY]
Wimpenny, —
Winston, Chas. H.
Winstrom, August [Philadelphia, PA]
Winter, T.D.
Winters, G.H. [Indianapolis, IN]
Winthrop, Dr. —
Wirth, —
Wiseman, F.C.
Wolfe, —
Wood, —
Wood, Edwin
Wooden, E.K.
Woodford, W.E.
Woodruff, Geo.
Woods, Ed
Wortham, T.R.
Worthen, —
Worthington, C.E.
Wortman, Dick
Wulff, Chas. G. (Also Wolff, Wolf)
Wyman, Daniel
Young, Edward [Port Huron, MI]
Young, Lark
Young, Robert

Women of '98-99

Ammann, Carrie O. (Mrs. Adolph). [Catskill, NY]
Barrett, Anna L. (Mrs. June J.)
Beatty, Isabel (Mrs. L.). [San

Jose, CA]
Bjornstad, Katie (Mrs. Martin B.)
Bonson, —. (Mrs. N.)
Brosman, Julia
Brown, —
Burr, —. (Mrs.)
Cameron, Minnie L. (Mrs. Fred J)
Cauvey (?), Barb
Claussen, —. (Mrs. —).
Daby, Eleanor (Mrs. —).
Darling, —. (Mrs.)
Date, — (Mrs. Fred)
Delander, Matilda (Mrs.)
Dodge, — (Mrs. —). Los Angeles, CA.
Doty, Eleanor F. (Nellie)
Dolling, Mrs. —
Dowling, — (Mrs. A.C.)
Flynn, — (Mrs. —)
Freiherz, — (Frau —)
Glaudman, Tress
Goss, — (Mrs. P. B.)
Hass, — (Mrs. —)
Helmers, — (Mrs. —)
Hosford, — (Mrs. O.J.)
Johnson, Rose (Mrs. Charles)
Kanitz, — (Mrs. Otto)
Kennedy, Agnes
Lyon, — (Mrs. L.)
McGee, Sarah F. (Mrs. W. J.)
Miller, —. (Mrs. —)
Moore, Lillian
O'Connell, Minnie
Peets, — (Mrs. E. L.)
Peets, — (Mrs. H. G.) California
Ransom, Fay
Ruppenthalt, Lena
Simmons, Mae
Smith, Rachel
Smith, —. (Mrs. Wm.)
Smith, M. (Mrs. Fred)
Syring, Minnie (Mrs. Herman)
von Gunther, Lena (Mrs. A.)
Wilson, Bessie L.

Selected Bibliography and Guide to Sources

The voices in this book come from a variety of published and unpublished records left by those who followed the Valdez gold rush trails of 1898 and 1899. Whenever possible, we identified the source in the text. Readers seeking additional information may find the author's name listed below. For published materials, we assume that readers wishing to pursue subjects further have access to detailed bibliographical information through their local libraries and databases such as GNOSIS on SLED. Many unpublished documents are preserved in the Crary Scrapbooks at The Anchorage Museum of History and Art. We refer to the Crary Scrapbooks by the abbreviation CS.

BOOKS: ABERCROMBIE, Capt. W.R., Alaska, 1899, Copper River Exploring Expedition (=Abercrombie, 1900); ABERCROMBIE, et.al. Compilation of Narratives of Exploration in Alaska (= Abercrombie, Compilation); ALLEN, H.T. Report in Abercrombie, Compilation; AUSTIN, Basil, The Diary of a Ninety-Eighter (= Austin); BABCOCK, Lt. Report in Abercrombie, 1900; BERTON, Pierre, The Klondike Fever; BRONSON, William. The Last Great Adventure; BROOKFIELD, Lt., Report in Glenn & Abercrombie; BROWN, Charley, Report in Abercrombie, 1900; Report in Glenn & Abercrombie; BROWN, Charles. The U.S. Army in Alaska; CASHMAN, Peter, Report in Abercrombie, 1900; CAVAGNOL, Joseph J., Postmarked Alaska, Chapter 15; GILLETTE, E. Report in Abercrombie, 1900; GLENN AND ABERCROMBIE, Reports of Explorations in the Territory of Alaska, 1898 (= Glenn & Abercrombie); HARRIS, A.C., The Klondike Gold Fields; HOLESKI, C.J. & M.C. HOLESKI, In Search of Gold: The Alaska Journals of Horace S. Conger, 1898-1899 (= Holeski for commentary; Conger for Journals); KETTLESON, Clara. The Golden Thread: A biography of Theodore Kettleson (1874-1968) as told by his wife; LOWE, P.G., Report in Glenn & Abercrombie; MARGESON, Charles, A., Experiences of Gold Hunters in Alaska (=Margeson); POWELL, Addison M., Trailing and Camping in Alaska (= Powell); Report in Abercrombie, 1900; PRESTON, G.H., Report in Glenn & Abercrombie; RAFFERTY, J.J., Report in Glenn & Abercrombie; REMINGTON, Charles H. (Copper River Joe), A Golden Cross (?) on the Trails from the Valdez Glacier

(=Remington); SCHRADER, F.C., Report and maps in Maps and Descriptions of Routes of Exploration in Alaska in 1898, with General Information Concerning the Territory; RICE, John, Report in Abercrombie, 1900; TOWNSEND, Dr. L.S., Report in Abercrombie, 1900; TOWNSEND, Peggy Jean, *The Alaska Gold Rush Letters of Leroy Stewart Townsend: 1898-1899.* Copyright © by Peggy Jean Townsend and Klondike Research, publication in press, due in 1997. WHARTON, DAVID B., The Alaska Gold Rush.

PERIODICALS: HAZELET, George C. Railroad Fever in Valdez, 1898-1907, Intro. and annotated by Elizabeth Towers, *Alaska History*; SWAN, Adam. Early Settlement of Valdez. *Pathfinder.*

NEWSPAPERS: The Valdez News is abbreviated as TVN.

ABERCROMBIE EXPEDITION 1899, *The Alaskan*, 11/11/1899; AMY, WINFIELD Scott, "Story of Founding of Valdez," in *The All-Alaska Review*, 1:3, 7/1915, p. 8; BRADY, Governor John. The Governor's Report, Mining. *The Alaskan*, 3/4/1898; CHESNA CREEK, legal problems: Mushers Arrive from Chesna.*TVN*. 1:10. 5/11/01. p. 1; Restraining Orders. Chesna Properties Tied up for the Summer. *TVN*. 1:11. 5/18/01. p. 1 Miners Meeting. Held on Slate Creek June 3rd. Petition to Judge Brown. *TVN*. 1:16. 6/22/01. p. 1; Restraining orders Issued. Upper Chisna Property Completely Tied Up. Judge Brown Opens Court at an unusual Hour in Order to Oblige Litigants. *TVN*. 1:19. 7/13/01, p. 4; COPPER RIVER ROUTE, "Report of Another Route to the Klondike," *San Francisco Chronicle*, 8/3/1897; "The Copper River Route," *Seattle-Post Intelligencer*, 8/8/1897; DEMPSEY's Mansion, Report from Chisna, *TVN* 2:15. 6/21/02; DOC TANNER, "Doc Tanner Lynched," *San Francisco Examiner*, 2/2/1898; "GOLD! GOLD!" *Seattle Post-Inteligencer*, 7/17/1897; GUITEAU, Louis "Death Trail of 98 over the Valdez Glacier," in *The Alaskana, II* , 9/1972 (= Guiteau, Death Trail); GUITEAU, Luther W., "Alaska Gold Rush of 1898." Serialized in Freeport, IL's *Journal-Standard,* 1928 (= Guiteau); HUMPHREY, Capt. Omar J., "New Way to the Klondike," *The Alaskan,* 9/25/1897; "Rushing for Copper River," *San Francisco Chronicle*, 10/21/97; "New Route to the Yukon Over American Soil," *Seattle Post-Intel-ligencer*, 11/18/1897; Observations on Copper River Area Rush, *The Alaskan*, 11/26/1898; INTOXICAT-ING LIQUORS. Executive order making intoxicating liquors illegal in Alaska, text reprinted in *The Alaskan*, 7/16/1898; *LA NINFA* DEPARTS, "News of the Ocean and Waterfront," *San Francisco Chronicle*, 11/11/1897; LOGAN PARTY, Report on deaths of the Logan party and B.A. Benson. *The Alaskan*, 3/18/1898; LOWE, P.G. Lt. Lowe and Men Overland from Valdes to Forty-mile City, *The Alaskan*, 11/26/1898; MAIL, possibility of mail between Valdez and Eagle City winter of 1899-1900, *The Alaskan*, 11/11/1899; MILLARD, B. F. The Chittyna Copper Belt. The True History of How it was Found. A Race to Interior.*TVN*. 1: 7. 4/20/01. p. 1; 1: 8. 4/27/01; p. 1; 1:9. 5/04/01; PACIFIC STEAM WHALING COM-PANY, (see also Copper River & Valdez entries), "Rushing for Copper River," *San Francisco Chron-icle*, 10/21/97; "News of the Ocean and Waterfront," *San Francisco Chronicle*, 11/14/1897; 11/30/1897; POWELL, Addison. "History of Copper River Coun-try," original source unknown In Crary Scrapbook, II, p. 204-205; POWELL Description of the [Copper River] Country, *The Alaskan*, 3/25/1899; PRINCE WILLIAM SOUND & COPPER RIVER AREA, Miners headed to Copper River, *The Alaskan*, 5/11/1895; New Copper Mines, *The Alaskan*, 8/28/97; QUARTZ CREEK MINING. 125 mining area in winter.*The Alaskan*, 3/18/99; RAILROADS (from Valdez), What is needed, TVN. 1:5. 4/6/01; Pro-posed railroads, TVN, 1:6. 4/16/01; Railroad story, TVN. 1:10. 5/11/01; No Way Business! TVN. 5/11/01; Capt. Healy's Big Scheme. TVN.1:16. 6/23/01; M.J. Heney, TVN. 1:16. 6/22/01; Railroad rumor, TVN. 1:19. 7/13/01; Capt. Healy, TVN. 1:21. 7/27/01; More Railroad men coming, TVN. 1:22. 8/03/01; Another rumor, TVN. 1:23. 8/10/01; Heney returned, TVN. 1:25. 8/24/01; More Talk of Railroad, TVN. 1:30. 9/28/01; Right of Way Granted, 1:34. 10/26/01; SCURVY in the Copper River Country, *The Alaskan*, 3/1/8/99; SWAN, Adam, "Over the Glacier in 1898, *The Valdez Miner*, date unknown in CSc, II; TELEGRAPH, Transporting Wire.*TVN*. 1: 6. 4/13/01; Government telegraph line ... *TVN*. 1:8. 4/27/01; By Telegraph: First messages. *TVN* 2:5. 4/12/02; Valdez-Eagle telegraph linked. *TVN* 2:24. 8/23/02; VALDEZ, Today and Yesterday. Valdez: Its Past Present and Future."*TVN,* 3/9/1901; VALDEZ TRAIL & COPPER RIVER COUNTRY, "New Way to the Klondike," *The Alaskan*, 9/25/1897; "To Pros-pect Copper River," *San Francisco Chronicle*, 8/30/1897; "Bound for Copper River," "Mines of Copper

River," *San Francisco Chronicle*, 9/12/1897; "New Route to the Yukon Over American Soil," *Seattle Post-Intelligencer*, 11/18/1897; 10/11/1897; "The Latest Map, Alaska," *San Francisco Chronicle,* 12/30/1897; "There is gold in Alaska all the way to the coast,"*The Alaskan,* 2/19/1898; The All-American Route to the Klondike, *The Alaskan,* 3/19/1898;

Diaries, Journals, Letters, Memoirs:

BENEDICT, Neal, The Valdes and Copper River Trail, Alaska, Alaska State Library; BOURKE, Jo-seph, Journal, The Valdez Museum and Historical Archives; BOURKE, Joseph, Scrapbook, The City of Valdez Archives; BOURKE, Joseph, Notes, Val-dez, 1898, CS; CRARY, Will H., Diary, CS; DEMP-SEY, Melvin, Diary 1898, CS, I:251; Diary, 1900, CS, I:199; FLEMING, Henry, A True Account of the Alaskan Adventures of Henry Fleming (Gold Rush 1900-1902), typescript compiled by Christina Coun-selor, 1976, Valdez Museum and Historical Ar-chives; HAZELET, George C., Journals, typed manu-script in Valdez Museum and Historical Archives; MOORE, Lillian, Letter, Valdez Museum and His-torical Archives; PEARSON, Bryan, Extracts from Journal of H. Bryan Pearson, CS, 1:185-187; TOWNSEND, Dr. Leroy, *The Alaska Gold Rush Letters of Leroy Stewart Townsend: 1898-1899.* Copyright © by Peggy Jean Townsend and Klondike Research, publication in press, due in 1997. TRE-LOAR, William E., Memoirs, Valdez Historical Society Collection. TUFFIN, Horace, The Diary of Horace Y. Tuffin in the Charlotte Hazelet Turtainen Collection, Valdez Museum and Historical Archives.

Mining Law: BROOKS, Alfred Hulse. Blazing Alas-ka's Trails. pp. 509-511; HARRIS, A.C. "Resume of Mining Laws," in Alaska and the Klondike Gold Fields, pp. 402-421 and INGERSOLL, ERNEST, Gold Fields of the Klondike and the Wonders of Alaska, (pp. 208-222) typify the type of handbook infomation prospectors would have had on mining laws; MARKS, PAULA MITCHELL, Precious Dust: The American Gold Rush Era: 1848-1900 provides historical background on miners' law and miners' courts; POWER OF ATTORNEY, problems with: Power of Attorney. May be Knocked out by Con-gress. *TVN*. 1:51. 2/22/02. p. 1; Power of Attorney.*TVN,* 2:21. 8/2/02, p. 2; Ivey Talks. citizens hear him speak on needed legislation. *TVN* 2:32 10/18/02, p.1; SHEPHERD, Thomas R., Placer Mining Law in Alaska, 1909. (Reprinted from the Yale Law Journal, May 1909) gives the best analysis of the law and its problems as applied to Alaska.

Mining Records are found at the Valdez Court House. Records used include Mining Locations Book G with parts of H & I (Prince William Sound=PWS), Prince William Sound Mining District 6/11/1898 to 5/9/1900; Mining Locations, Prince William Sound, Book 2; Mining Locations, Chisna District, Books J & F in Book 3 3/1898 to 3/1902, (contains some Prince William Sound); Kotsena Mining District Records; Tonsena Mining District Records (includes some McCarthy District claims); Quartz Creek Mining District Records; Valdez Mining Records (=VDMR), Book 1 (includes some Chesna district claims); DEEDS, Book A. 2/1/1901 to 7/25/1901.

Miscellaneous: ANNUAL REPORT (for sub-port of Valdez), 1905, E.B. Spiers, Deputy Collector of Customs, (=CS), I:267-288; BOURKE, Lester, Letter about Joseph Bourke, Valdez Museum and Historical Archives; CRARY PHOTOGRAPHIC SCRAPBOOKS, The Anchorage Museum of History and Art; CORONER'S REPORT for LOGAN PARTY, CS, I; CORONER'S REPORT Depicting Death of Joseph Fournier et al on Valdez Glacier, 1898, Valdez Museum and Historical Archives; HUBBARD, Charles Goodyear, Diary, University of Alaska, Anchorage Archives; KERTCHEM, J.F., Letter to Lands Commissioner B. Hermann, VMDR, Vol. 1, p. 192; MEALS, Frank, Letter to Owen and Nancy Meals, 5/12/1966; MEMBERSHIP RECORDS, Christian Endeavor Society, Valdes, Alaska, Valdez Historical Society; PIONEERS OF ALASKA, IGLOO No. 7, VALDEZ. Information on ages and places of birth, University of Alaska Archives, Fairbanks, Alaska, Pioneer Records; PROSPECTUS of the Connecticut and Alaska Mining and Trading Association, CS, pp.112-114; ROSENTHAL, F.W., Presentation Speech, 1/1908, CS, 1, p. 215.

Valdez township and townsite, an abbreviated version of the council MINUTES occurs in the CS, 1, pp. 76a-89. Interspersed are the only known extant minutes of the citizens' meetings; complete minutes of the early township and townsite meetings occur in The Valdez City Archives. Records of Valdez townsite lot claims for 1898-1899, occur in INDEX LOTS AND LANDS, BOOK I, Valdez Recording District, located at Alaska Court House Office in Valdez.

Guide to Sources by Chapter Sections: CAPITALIZED WORDS refer back to bibliographic entries.

Chapter 1: The Gold Fever of '98, pp. 1-17

The Steamer Valencia arrives in Port Valdez: The arrival of the *Valencia* is based on descriptions from BOURKE, (Journals), HAZELET, (Diaries, 3/13/98), TRELOAR, (Memoirs). Treloar, a first class passenger on the *Valencia*, says she carried 640 passengers, while Bourke in second class says 580. We used Treloar's figures because he probably had better access to the ship's officers who kept these records. Bourke may be recording only the number crowded together in second class steerage. Treloar (pp. 13-14) gives a vivid account of the revolt on the *Valencia*.

The Historical Setting: For panic of 1893 see BRONSON, pp. 1-2.

The Klondike and the Copper River Gold Rush: See Berton (p. 115) for use of the word "Klondike." The bogus map appeared in the *San Francisco Chronicle*, 12/30/1897. See COPPER RIVER ROUTE and VALDEZ GLACIER TRAIL & COPPER RIVER COUNTRY in bibliographic section.

Picking the Best Route and Destination: The Newspapers: The HOLESKIS' introduction provides excellent background coverage on the hyping of the Copper River Area and Valdez Glacier Route. **Government Reports:** ABERCROMBIE'S 1884 report occurs in Abercrombie, et. al. *Compilations of Narratives,* pp. 383-391; ALLEN'S 1885 report occurs in the same book, pp. 409-494. Allen's three-man team of himself, Sgt. C. Robertson and Prt. F. Fickett were a hand-picked, elite team.

The Valdez Glacier Diarists: Pierre BERTON in *Klondike Fever* estimates that not more than .5% of the original 3500 to 4000 prospectors made it to the Yukon (p. 216). SCHRADER also notes the Copper River as the prospectors' intended destination but adds, "Many hoped at the same time to proceed by way of the prospector's All-American route into the gold disticts of the Upper Yukon (p. 368)." Some prospectors ended up in the Yukon by default because prospects in the Copper River did not pan out. California prospector and diarist TRELOAR after a futile summer in the Copper River Country proceded on to Forty-Mile, Dawson, and ultimately the Nome gold fields. However, most felt they woud find gold in the Copper River Country or on the route to the Yukon.

Chapter 2: The Gold Rushers Arrive at Port Valdes, pp. 18-26

Early Arrivals at Orca: For sources of information in this section see AMY, LA NINFA, HUB-BARD, and SWAN. HUBBARD's Diary describes conditions in Orca and the route up the Copper River. POWELL's History provides information on Jackson, while BOURKE, SWAN and MARGESON are sources for Olsen and Beyer. The story of Omelia appears in MARGESON, pp. 60-61.

Early Prospecting in Prince William Sound: Sources for the permanent residents are the Mining Locations records and Pioneers of Alaska, Igloo No. 7: For A. Lind, and O. Carlson see Book G. p. 466; for Tom Olsen, John Ellis, Louie Thorstensen, Nils John Jacobsen (Norway), Jack Shepard, and Larsen see PWS. Bk. 1, p. 33. Tom Olsen's filing here at Virgin Bay contradicts Bourke's statement that Olsen was unable to file because he was not a U.S. citizen (BOURKE, Notes). For Bald-headed Chris see Book G, pp 300-301. W.J. Busby filed his first claim on Middleton Island in 1893 (Book G), and served as the Recorder for the Prince William Sound Mining District for many years, cf. PWS Book 1, 2. Chris Christiansen (Norway) attended the 1897 PWS Mining District meeting, BWS, Bk 1; Beyer (Germany) located the Olga Mine at Latouche in 1897 but did not record. Sources for the summer miners from the Sunrise district on Turnagain Arm in Cook Inlet come from Title Bonds in the Mining Location Records, thus Gladhaugh (Book G, p. 428) Sanford J. Mills (Book G, p. 437), W.E. Hunt, (Book G, p. 437). W.E. Ripstein recorded one of the first claims at Galena Bay in 1896 (Book G, p. 57, 60). See also for W.E. Ripstein, A.K. Beatson, M.O. Gladhaugh, W.B. Heidorn, Pete Jackson (Sweden), and W.E. Hunt, see PWS, Bk. 1, pp. 23-25; William E. Ripstein is undoubtedly the prospector from Sunrise called Billy Ribbstein that Tapper Bayles reports crossed Valdez Glacier in 1897, *The Alaskan*, 3/19/98.

Gentlemen summer visitors included the McNair party of A.K. Delaney (Juneau, AK), C. H. Anderson (Ontario and San Francisco, CA), E. E. Jones (Sitka, AK), L.L. Williams (Juneau, AK), E. de Groff (Sitka, AK), W.N. McNair (Sitka, AK), W.D. McNair (Sitka, AK), PWS Mining District Records, Bk.1, p. 26; Book G, pp. 324ff., For McNairs and de Groff's occupations, see *The Alaskan*, 4/23/98. In the summer of 1897, the McNair party recorded claims in Landlocked Bay and Solomon Gulch. For F.C. Lawrence from Ashcroft, B.C. Canada see Book G. pp. 319-324, 437.

The first stampeders arrive in Port Valdez: Major sources include SWAN, AMY; KETTLESON, pp. 96-97. For the *Wolcott* and *Behring Sea* see the *San Francisco Chronicle*, 10/21/1887 and 11/11/

1897. The best source for the Pacific Steam Whaling Company's Seattle and San Francisco media campaign is HOLESKI AND HOLESKI.

Establishing the Law and Order Trail: The account of Doc Tanner primarily follows Charles HUBBARD'S report to the *San Francisco Examiner, 2/2/1898*. Hubbard arrived in Valdes the day after the hanging and wrote his detailed account at that time. AMY, who participated in the events, wrote about it 17 years later for the *All Alaska Review,* Names, dates, reported speech, and interpretations differ. Both POWELL in *Trailing* and REMINGTON in *Golden Cross (?)* give accounts. Powell's is based on Amy's in detail but portrays Doc Tanner as the achetypal American poor man, while Remington, a staunch prohibitionist, uses the episode as an illustration of the powers of alcohol on a weak man. Hubbard, however, says of Tanner "tobacco he never used and he was never seen to use any kind of liquor." The one fault Hubbard notes was that Tanner "was given to boasting, remarking to one of the party on the way here that he could draw a gun as quickly as any man in America." Hubbard also places part of the blame on the Pacific Steam Whaling Company's false advertising.

For information on Mining Law see the Bibliographic section.

The Rush Begins: Reference to the exploratory party that crossed Valdez Glacier in January comes from the Valdez News Miner, 6/18/1935. The party encountered a severe blizzard on the return trip and saved themselves by hiding in crevasses. Greenig took a month to recover from the near catastrophic crossing.

Spring 1898 — The Stampede: Benedict (pp. 15-16) describes the unloading.

"Copper City" was the earliest name given to Valdez. AMY writing after the great copper strikes, claims that he and many of the early arrivals came in search of copper not gold (p.8). Examples of use: 4/4/98 in Prince William Sound Mining Distict, Book 1, p. 47; CRARY, Diary, 4/22/1898; Photo from Bourke Scrapbook.

"Camp Valdes" BENEDICT, text and photo p. 17; photo reprinted on page 25. "Port Valdes," "Valdes" and "Valdez" are discussed in Chapter 3. Estimates of the number of people in camp and on the glacier varied. CONGER reported 900 on 3/8/1898, while HAZELET estimates 700 on 3/13/1898. TOWNSEND reports the loss of the ice wharf in his letter of 3/20/1898. BOURKE (Journal, p. 3) mentions Louis Thorstensen's lightering operation. GUITEAU (Diary, 3/8/98) gives the wages. REM-

INGTON admired Middleton Island Smith (p. 5). LESLIE'S WEEKLY (12/8/1898) mentions Jackson giving up his cannery job to become the mail carrier. Ripstein's power-of-attorney agreements are in PWS Mining District Records, Bk. 1, pp. 57 ff. CONGER gives the temperature at the summit in his 4/21/1898 entry.

Chapter 3. The Founding of Valdez, pp.27-37

The Merchants Arrive: MARGESON gives a good account of the difficulties the Connecticut Company had in building their store, pp. 65-67. The store was located in Lot 11, Block 10 and fronted on Keystone Ave. CRARY'S Diary mentions Gray's Hotel on 6/11/1898. This is the earliest mention of a hotel in Valdez. Smith W. Gray chaired the controversial Township Committee until the citizens disbanded it (MINUTES, 4/26/1898; 5/11/1898.) In May 1898, he claimed a townsite Lot 4 in Block 8 on the corner of McKinley and Glacier (Lots and Lands Book I, p. 3). There are photographs of the Valdez Hotel taken in the summer of 1898, but whether this was owned by Gray is unknown. A photograph of the photography shop occurs in the Crary Photographic Scrapbooks.

The Townsite Rush: For Swanport, see SWAN, AMY; Copper City, see SWAN, AMY; for Swan's townsite next to Copper City, see SWAN; for Hangtown, see SWAN, AMY; for the Portland Group's site see KERTCHEM LETTER in VMDR, p. 192 and placer claims on pp. 31-32; for Keystone Company's townsite see TOWNSEND letters. Reference to the J. M. Hayse townsite occurs in the VMDR, p. 36. The *Pathfinder* article on the "Early Settlement of Valdez," states that "Mr. Swan made up his mind to stake out a townsite so he called a meeting and the 80 people present adopted a code of rules and regulations under which each man or woman could possess a lot." The minutes of both the township and citizens' committees contain no references to Adam Swan. This does not preclude his playing a role behind the scenes but of this there is no record. See also VALDEZ,*TVN,* 3/9/1901. The citizens' concerns are illustrated by the first ordinances they passed (MINUTES, 5/16/1898; 5/19/1898; 6/3/1898; 6/20/1898.)

Minutes from 4/23/1898 to 5/10/1898 cover the period of controversy. Citizen meetings continued until the first Valdez City Council was installed on 9/4/1901. For information on flooding, see ABERCROMBIE, 1900; MINUTES 1/22/1901; *TVN,* 5/11/1901; 6/8/01; 7/13/1901; 3/29/02; 10/04/02. Photos in Crary Photographic Scrapbooks.

Old-timers: W.J. Busby. See Bibliography, VALDEZ for location of records on townsite lots. The last Townsite Committee meeting Busby attended was June 17, 1898 (MINUTES 6/17/1898). Ordinance No. 12 pertaining to the location of townsite records in the absence of the recorder and deputy recorder passed the townsite committee on 7/18/1898 MINUTES; The Recorders' position was declared vacant at the Citizens' meeting on 10/10/1898 and J.M. Riggins was elected Recorder (MINUTES, CS, 10/10/1898).

The Valdez Mining District was established on 3/12/1898 in B & Co.'s tent. S.W. Gray received 5 votes for Recorder and J. F. Kertchem 7. If everyone attending voted, then twelve people attended the meeting, which fulfilled the legal requirement for a quorum. No information is available on whether the meeting was advertised for ten days in advance. For claims in Solomon Gulch and Basin area see Caryl, VMDR, pp. 48-50; other claims were located in the Mineral Creek Area and Abercrombie Gulch. For Fenske's claims on Glacier Island, 8/15/98 see PWS, Book 2. p. 46,47; S.V. Gray filed on Glacier Island on 8/8/1898, PWS, Book 2, p. 33, 98. For minutes of the 1/16/1899 meeting establishing the Valdez Recording District see VMDR, pp. 16-17; For W.E. Hunt's activities, see Mining Locations, Bk. G. pp. 437, 701-704; Lawrence, Bk. G, p. 428; Ripstein, PWS, Bk. 1, p. 76; Christiansen, PWS, Bk. 2, pp. 223-4.

New-comers: Melvin Dempsey. Sources include DEMPSEY, Diary; BOURKE, Notes; VMDR, pp. 13-14; MINUTES, 5/9/1898; 5/19/1898; 10/10/98; For Dempsey's ancestry see MEALS, Letter.

How do you pronounce the Town's Name? BOURKE, Journal, gives the prospector's mailing address. Note also the spelling of Valdes on the early maps. In copying the Valdez townsite minutes, Crary only sporadically preserved the "s" and "z" spelling, see original MINUTES in City of Valdez Archives. TUFFIN, 3/8/1900. *The Alaskan*, 3/19/98.

Chapter 4: Crossing Valdez Glacier, pp. 38-66

Surmounting the Terminus: For the three men who perished in the avalanche, see HOLESKI, pp. 76-77; for the second and third benches, see BENEDICT, p. 22 and MARGESON, p. 76;

A Final View of Civilization: See BENEDICT (p. 24) for description of method of surmounting the Third Bench; BOURKE (Journal, p. 24) remarks that tents and gear were the last items to be moved; AUSTIN (p. 29) describes the push to the base of the Fourth Bench.

Five-mile Camp: We discuss Abercrombie's

one interference in the illegal trade in alcohol on page 96.

Animals on the Trail: Information on the use of horses, effect on trail and future plans to obtain them come from AUSTIN, BOURKE, MARGESON, BROOKFIELD, AND PRESTON. BROWN mentions that Bruce, H. Smith, Oscar Peterson, Malmgren, & Leveross arrived in Valdez Nov. 7, 1898 from interior to go outside to get pack animals. They planned to return in February, BROWN in Glenn and Abercrombie, p. 623.

Women on the Trail: Hazelet estimates there were 30 to 40 women on the trail; we have found records for 44 women, but some of these women, such as Mrs. Goss, probably remained in Valdez. BOURKE (Scrapbook) gives the names of 11 women who crossed the glacier. HAZELET, MARGESON, AND BENEDICT all mention the well-dressed redhead. PRESTON (p. 601) mentions Mrs. Dowling's nursing Private Tully (also Tolley). Concerning Mrs. Dowling winning the rifle shoot, CONGER wrote, ". . . they were having a rifle shoot. . . Five ladies took a hand in the shooting. Mrs. Dowling won first money. (9/10/1898)." The map on page 80 shows the location of Dowling Peak. Although the diarists generally praised the virtue of women on the trail, DEMPSEY records, "Mrs. Darling and Doc Dinsmore left in open boat. Said to be eloping (DEMPSEY, 11/5/1898)."

12-Mile Camp: BROOKFIELD describes hand pulling up the fourth bench;

Lost on the Glacier: For GUITEAU'S confusion over his location see Diary, May 22, 23, 24, and 28th, 1898; MARGESON (p. 57) describes sending a person to consult with Omelia.

Summit Camp and the Great Storm: BOURKE (Journal, p. 27) gives a good account of conditions at the camp on April 30th.

Avalanche: MARGESON (p. 89) and TRELOAR (pp. 27-28) describe the avalanche, depth of the snow and response; BOURKE, (Journal, p. 31-35) describes his experiences in rescuing Joseph Thiery; the CORNER'S REPORT provides information on the conditions, rescue attempt, the men who participated, and on Antwerp and Fournier; MARGESON (p. 102) and BENEDICT (p. 169) tell the story of Shorty's dog. Also POWELL and REMINGTON.

Finally the Cavalry Arrives — Late: See INTOXICATING LIQUORS under Newspapers. MOORE, Letter, p. 2 mentions the absence of blankets. MOORE, SCHRADER, KOEHLER AND ABERCROMBIE all describe the condition of the horses. CASHMAN describes his trip to retrieve the horses and their fate.

Chapter 5: Coming into the Country — Down the Klutina. pp. 67-85.

Descending into the Promised Land: GUITEAU, Diary (5/1/1898) reports that West was unable to push his trail through. For HAZELET's account see Journal, 5/1/1898.

The Upper Klutina: Both BENEDICT (p. 64) and MARGESON (pp. 126-128) report Kelley's death.

Klutina Lake: For the confusion between Klutina Lake and River and Abercrombie's mythical Lake Margaret and the Tasnuna River, see RAFFERTY (p. 615), CONGER, 6/7-9/1898; and CRARY, Diary, 6/8/1898. MARGESON (p. 143) describes Jackson's shrill whistle when he arrived with the mail. CRARY (7/21/1898) and RAFFERTY (p. 601) describe finding caches left by earlier white explorers. Both BOURKE (Journal, p. 79-80) and Holeski (p. 38) discuss the price and problems of mail delivery. TOWNSEND gives the total for his mai delivery in his letter of 10/10/1898. Problems between the Orca and Valdez post masters are discussed in GLENN & ABERCROMBIE, p. 368, while Klutina City is described on p. 358.

The Indians: For Indian Charley's advice see HAZELET, 11/21/1898 and POWELL, *Trailing*, p. 231. For the Indian's use of discarded clothing, see MARGESON (p. 136).

The Rapids of the Lower Klutina: ABERCROMBIE, in Glenn and Abercrombie, p. 311; both BOURKE (p. 49) and HOLESKI (p. 106) report paid river guides.

Gold at Quartz Creek: Information in this section comes from a study of the Quartz Creek Mining Records. Phil Trevedt and Fred Corless' (also Fred Corlis) claims on the east side of Quartz Creek were located on 8/15/98, see the Quartz Creek Records, pp. 128-129. Fred Corlis was from Carpenteria and joined Treloar's group there (TRELOAR, p. 1). Treloar writes that on June 6th "we had a little trouble with Corlis. . . to keep peace and save him from a miners' meeting, we divided and gave him his share to get rid of him, for we would have no rows or trouble in camp if there was a way to avoid it. (TRELOAR, p. 33)." On p. 34, Treloar describes a knife fight and says "This was the cause of Corlis being let out of our party."

The brothers, Joe Remington and Grant Remmington, consistently spelled their last names differently, see Quartz Creek Records pp. 31, 55.

Copper Center: BOURKE, Journal, p. 65 describes the development of Copper Center.

Entertainment on the Trail: MARGESON, REMINGTON, POWELL, AND MOORE all describe H. E. F. King's musical abilities.

Chapter 6: Winter Exodus. pp. 87-97.

Scurvy, Scourge of the Prospectors: We wish again to acknowledge are indebtedness to Peggy Jean Townsend and Klondike Research, for permission to use and quote from *The Alaska Gold Rush Letters of Leroy Stewart Townsend: 1898-1899.* Copyright © by Peggy Jean Townsend and Klondike Research, publication in press, due in 1997. See TOWNSEND, Report. on misdiagnosis by other doctors; for death from scurvy in May, see CONGER (5/26/1898); for foods rich in vitamin C as a treatment, see AUSTIN, TOWNSEND; for belief in curative power of fresh meat, see GUITEAU, Diary and CONGER, 4/10/1899 to 8/20/1899. Conger gives a most detailed account of the effects of scurvy on the victiim's extremeties Conger's entry of 6/5/1899 shows the difficult choices many parties made: "The fiends incarnate, every last one of them, have gone out for a 3 weeks' prospect and left me alone, helpless, with my leg black to the knee. It is help yourself or die now." Because of similar experiences, contracts began including provisions for the illness or injury. A one year partnership between R.F. McClellan, E.A. Gates, H.E. Gates, H. H. Fitch, J.E. Hamlin, W.S. Amy, D.S. Kain, J. Sweeny, J.H. Smith & C. Warner states: "that in case any of the parties hereto shall bcome ill, or are otherwise incapacitated from pursing said business, then the other parties hereto shall at the joint expense of all the parties hereto care for such party, ill or incapacitatd, and he shall receive his share of the joint property, during the lifetime of this agreement." (DEEDS, Bk. A, p. 163).

TOWNSEND and REMINGTON both comment on the government's inability to provide relief during the winter.

Abercrombie's Reports: BROWN (Abercrombie, 1900, pp. 37-38) mentions the party of 25; For routine use of Valdez Glacier by prospectors going to and from the interior see, *TVN:* 4/6/01; 4/20/01; 3/8/02; 3/15/02; 3/29/02; 5/10/02; 2/28/03. In 1900, HAZELET, using the government trail, arrived at his Chestochina claims almost two months after those who went via the glacier, see Diary, 5/28/1900 and 8/5/1900. This contibuted to his ultimate business failure as a miner.

Chapter 7: A Glacier-free Route to Alaska's Interior. pp. 98-117.

Military Route and Road 1898-1899: ABERCROMBIE mentions the Russian trail, Glenn & Abercrombie, pp. 564, 586;

Corporal Heiden wins the Day: Hazelet mistakenly credits Schrader with finding the route through Keystone Canyon. See HAZELET, Railroad Fever in Valdez, p. 30.

Starting the Trans-Alaska Military Trail: For information on the great copper race, see MILLARD, True History; ROHN (pp. 113-4) reports McClellan, Amy and Millard got through Keystone Canyon before Rice, Young, Downey & Holman parties did on June 19th and are building trail at Tsaina when he meets McClellan, Amy, Millard & others on Tsaina; Rohn travels with Millard and Warner up Kotsena to copper area; PEARSON arrives at Millard's camp, 4/22/1899.

Prospecting Results in 1899: Hazelet stirred up considerable controversy by using the power-of-attorney to stake claims for people back home. Local prospectors felt this was unfair and illegal; they jumped his claims, see HAZELET, 8/5/1900; 8/24/1900; 9/2/1900; 9/20/1900; 9/28/1900; Hazelet writes "Old Dempsey is the remote cause of all this and he shall yet have the district in disgrace (8/24/1900)." See also, CHESNA CREEK, legal problems and POWER OF ATTORNEY. FOR HAZELET'S bankruptcy, see Diary, 8/5/1902; See DEMPSEY'S MANSION.

Finding a Railway Route: see RAILROADS under newspapers.

Ft. Liscum: see BROWN, Charles. The U.S. Army in Alaska.

The All-Alaska Telegraph Line: See TELEGRAPH under Newspapers above and BROWN, C. The U.S. Army in Alaska.

Index